NONLINEAR PHYSICAL SCIENCE

NONLINEAR PHYSICAL SCIENCE

Nonlinear Physical Science focuses on recent advances of fundamental theories and principles, analytical and symbolic approaches, as well as computational techniques in nonlinear physical science and nonlinear mathematics with engineering applications.

Topics of interest in *Nonlinear Physical Science* include but are not limited to:

- New findings and discoveries in nonlinear physics and mathematics
- Nonlinearity, complexity and mathematical structures in nonlinear physics
- Nonlinear phenomena and observations in nature and engineering
- Computational methods and theories in complex systems
- Lie group analysis, new theories and principles in mathematical modeling
- Stability, bifurcation, chaos and fractals in physical science and engineering
- Nonlinear chemical and biological physics
- Discontinuity, synchronization and natural complexity in the physical sciences

SERIES EDITORS

INTERNATIONAL ADVISORY BOARD

Ivo Petráš

Fractional-Order Nonlinear Systems

Modeling, Analysis and Simulation

With 119 figures

高等教育出版社
HIGHER EDUCATION PRESS

Springer

Author

Ivo Petráš
Technical University of Košice, Faculty of BERG
Institute of Control and Information of Production Processes, B. Němcovej 3
042 00 Košice, Slovak Republic
Email: ivo.petras@tuke.sk

ISSN 1867-8440 e-ISSN 1867-8459

Nonlinear Physical Science

ISBN 978-7-04-031534-9

Higher Education Press, Beijing

ISBN 978-3-642-18100-9 ISBN 978-3-642-18101-6 (eBook)

Springer Heidelberg Dordrecht London New York

Library of Congress Control Number: 2011920976

Printed on acid-free paper

Springer is part of Springer Science+Business Media (www.springer.com)

This book is dedicated to my entire family, my parents Štefan and Erika, and my sister Petra for their understanding and support.

Preface

The aim of the book is to present a survey of a new class of chaotic systems, the so-called fractional-order chaotic systems. This book can also be used as a textbook for courses related to nonlinear systems, fractional-order systems, etc. The book is suitable for advanced undergraduate and graduate students. It is a sort of a guide to fractional-order chaotic systems that features material from original research papers, including the author's own studies. The book is organized as follows:

Chapter 1 is a brief introduction to fractional-order chaotic systems.

Chapter 2 provides fundamentals of fractional calculus, its properties and integral transfer methods. Three well-known definitions of fractional derivatives/integrals and methods for their numerical approximation are presented.

Chapter 3 includes a presentation of fractional-order systems, their description and properties. Fractional linear time-invariant (LTI), nonlinear systems, and fractional-order controllers are considered.

Chapter 4 is devoted to stability of the fractional-order (LTI and nonlinear) systems. The stability of interval fractional-order system is also investigated.

Chapter 5 contains a survey of various fractional-order chaotic systems with the total order less than three. The well-known systems such as, for example, Chua's oscillator, Lorenz's system, Rössler's system, Duffing's system, and some other systems, for instance, Volta's system, are analyzed as well.

Chapter 6 begins with the introduction to control strategies of the fractional-order chaotic systems. Three general approaches: feed-back control, sliding mode control, and synchronization, are described. Other strategies are mentioned and discussed as well.

Chapter 7 concludes this book by some additional remarks.

Appendix A lists Matlab functions used for simulation of the fractional-order chaotic systems described in Chapter 5.

Appendix B lists a Laplace transform and inverse Laplace transform table of the functions used in fractional calculus.

Košice, April, 2010 *Ivo Petráš*

Acknowledgements

Besides new results, this book also presents results which have been already published by the author in several journal papers and conference articles during last almost 10 years. Because a portion of the material from these publications is reused in this book, the copyright permission was granted from several publishers.

Acknowledgement is given to Springer Science+Business Media for kind permission to reproduce a portion of the material from the following papers:

- Petráš I., Chaos in the fractional-order Volta's system: modeling and simulation, *Nonlinear Dynamics*, vol. 57, no. 1–2, 2009, pp. 157–170, DOI: 10.1007/s11071-008-9429-0.

Acknowledgement is given to the Institute of Electrical and Electronic Engineers (IEEE) for permission to reprint portions of the material from the following papers:

- © [2009] IEEE. Petráš I., Chen Y. Q. and Coopmans C., Fractional-order memristive systems, *Proc. of the IEEE Conference on Emerging Technologies & Factory Automation, ETFA 2009*, 22–25 Sept., 2009, Palma de Mallorca, Spain, DOI: 10.1109/ETFA.2009.5347142.
- © [2009] IEEE. Petráš I. and Bednárová D., Fractional-order chaotic systems, *Proc. of the IEEE Conference on Emerging Technologies & Factory Automation, ETFA 2009*, 22–25 Sept., 2009, Palma de Mallorca, Spain, DOI: 10.1109/ETFA.2009.5347112.
- © [2009] IEEE. Chen Y. Q., Petráš I. and Xue D., Fractional order control - A tutorial, *Proc. of the American Control Conference, ACC 2009.*, 10–12 June, 2009, St. Louis, USA, pp. 1397–1411, DOI: 10.1109/ACC.2009.5160719.
- © [2006] IEEE. Petráš I., A Note on the Fractional-Order Cellular Neural Networks, *Proc. of the International Joint Conference on Neural Networks*, 16–21 July, 2006, Vancouver, Canada, pp.1021–1024, DOI: 10.1109/IJCNN.2006.246798.
- © [2004] IEEE. Petráš I., Chen Y. Q., Vinagre B. M. and Podlubny, I., Stability of linear time invariant systems with interval fractional orders and interval coefficients, *Proc. of the Second IEEE International Conference on Computational Cybernetics, ICCC 2004*, Aug 30 – Sep 1, 2004, Vienna, Austria, pp. 341–346, DOI: 10.1109/ICCCYB.2004.1437745

Acknowledgement is given to the American Society of Mechanical Engineers (ASME) for permission to reprint portions of the material from the following paper:

- Coopmans C., Petráš I. and Chen Y. Q., Analogue fractional-order generalized memristive devices, *Proc. of the ASME 2009: International Design Engineering Technical Conferences & Computers and Information in Engineering Conference*, Aug 30 – Sep 2, 2009, San Diego, USA, DETC2009-86861.

Acknowledgement is given to Elsevier for permission to reprint portions of the material from the following papers:

- Petráš I., A note on the fractional-order Volta's system, *Communications in Nonlinear Science and Numerical Simulation*, vol. 15, no. 2, 2010, pp. 384–393, DOI: 10.1016/j.cnsns.2009.04.009.
- Petráš I., A note on the fractional-order Chua's system, *Chaos, Solitons & Fractals*, vol. 38, no. 1, 2008, pp. 140–147, DOI: 10.1016/j.chaos.2006.10.054.

Acknowledgement is also given to the copyright holder Diogenes Co. for permission to reproduce material from the following paper:

- Petráš I., Stability of fractional order systems with rational orders: A survey, *Fractional Calculus & Applied Analysis*, vol. 12, no. 3, 2009, pp. 269–298.

The author is grateful to the following people for their support, help and fruitful discussions:

Igor Podlubny, Ľubomir Dorčák, Imrich Koštial, Ján Terpák, and Gabriel Weiss (Technical University of Košice, Slovakia), YangQuan Chen (Utah State University in Logan, USA), Paul O'Leary (Montanuniversitat of Leoben, Austria), Blas M. Vinagre (University of Extremadura in Badajoz, Spain), Richard Magin (University of Illinois at Chicago, USA), Virginia Kiryakova (Institute of Mathematics & Informatics, Bulgarian Academy of Sciences, Sofia, Bulgaria), Riccardo Caponetto, Giovanni Dongola, and Luigi Fortuna (University of Catania, Italy).

I appreciate greatly Albert C. J. Luo (Southern Illinois University, Edwardsville, USA), the series editor of Nonlinear Physical Science co-published by Higher Education Press and Springer, for invitation and encouragement to write this book. I would like to acknowledge the help of Ms. Liping Wang, the publishing editor of Higher Education Press for professional comments, recommendations and help through at the review and production process with careful assistance.

I would like to express my thanks to Dr. Ladislav Pivka for improving the English translation.

I am also thankful to MathWorks, Inc. for Matlab and Dr. Duatre Valerio (Technical University of Lisbon, Portugal) for his excellent Matlab toolbox – *ninteger*.

Last but not least, I am also thankful to my entire family for their understanding and support.

Contents

Acronyms

A list of symbols and abbreviations used in the book:

${}_aD_t^\alpha$	fractional-order derivative/integral
$\mathbf{J}$	Jacobian matrix
T_{sim}	simulation time
L_m	memory length
T	sampling period
h	time step of calculation
$\mathbf{I}$	identity matrix
det	determinant
eig	eigenvalue
arg	argument of complex number
E^*	equilibrium point
Γ	Euler's gamma function
$E_{\alpha,\beta}(z)$	Mittag-Leffler function
$[.]$	integer part
B_p	breakpoint
$sign$	signum
$\approx$	approximation

GL	Grünwald-Letnikov definition
RL	Riemann-Liouville definition
PSE	Power Series Expansion
CFE	Continued Fraction Expansion
FIR	Finite Impulse response
IIR	Infinite Impulse Response
ORA	Oustaloup Recursive Approximation
PID	Proportional-Integral-Derivative
FOC	Fractional-Order Controller
CRONE	Commande Robuste Ordre Non Entier
TID	Tilt-Integral-Derivative
CPE	Constant Phase Element
LTI	Linear Time Invariant
FODE	Fractional-Order Differential Equation
FOS	Fractional-Order System
FOLTI	Fractional-Order Linear Time Invariant
BIBO	Bounded-Input Bounded-Output
LMI	Linear Matrix Inequality
FDEG	Fractional Degree
LCM	Least Common Multiple
VPO	Van der Pol
FrVPO	Fractional Van der Pol
CNN	Cellular Neural Network
LE	Lyapunov Exponent
MS	Memristive System
NMR	Nuclear Magnetic Resonace

Chapter 1
Introduction

Fractional calculus is a topic being more than 300 years old. The idea of fractional calculus has been known since the regular calculus, with the first reference probably being associated with Leibniz and L'Hospital in 1695 where half-order derivative was mentioned. In a correspondence between Johann Bernoulli and Leibniz in 1695, Leibniz mentioned the derivative of general order. In 1730 the subject of fractional calculus did not escape Euler's attention. J. L. Lagrange in 1772 contributed to fractional calculus indirectly, when he developed the law of exponents for differential operators. In 1812, P. S. Laplace defined the fractional derivative by means of integral and in 1819 S. F. Lacroix mentioned a derivative of arbitrary order in his 700-page long text, followed by J. B. J. Fourier in 1822, who mentioned the derivative of arbitrary order. The first use of fractional operations was made by N. H. Abel in 1823 in the solution of tautochrome problem. J. Liouville made the first major study of fractional calculus in 1832, where he applied his definitions to problems in theory. In 1867, A. K. Grünwald worked on the fractional operations. G. F. B. Riemann developed the theory of fractional integration during his school days and published his paper in 1892. A. V. Letnikov wrote several papers on this topic from 1868 to 1872. Oliver Heaviside published a collection of papers in 1892, where he showed the so-called Heaviside operational calculus concerned with linear generalized operators. In the period of 1900 to 1970 the principal contributors to the subject of fractional calculus were, for example, H. H. Hardy, S. Samko, H. Weyl, M. Riesz, S. Blair, etc. From 1970 to the present, they are for instance J. Spanier, K. B. Oldham, B. Ross, K. Nishimoto, O. Marichev, A. Kilbas, H. M. Srivastava, R. Bagley, K. S. Miller, M. Caputo, I. Podlubny, and many others (Cafagna, 2007; Miller and Ross, 1993).

At present, the number of applications of fractional calculus rapidly grows. These mathematical phenomena allow us to describe and model a real object more accurately than the classical "integer" methods. The real objects are generally fractional (Nakagava and Sorimachi, 1992; Oustaloup, 1995; Podlubny, 1999a; Westerlund, 2002), however, for many of them, the fractionality is very low. A typical example of a non-integer (fractional) order system is the voltage-current relation of a semi-infinite lossy transmission line (Wang, 1987) or diffusion of heat through a semi-infinite solid, where the heat flow is equal to the half-derivative of the temperature (Podlubny, 1999a).

The main reason for using integer-order models was the absence of solution methods for fractional differential equations. At present, there are many methods for approximation of the fractional derivative and integral, and fractional calculus can be easily used in wide areas of applications. Fractional-order calculus has played an important role in physics (Parada et al., 2007; Torvik and Bagley, 1984), electrical engineering (electrical circuits theory and fractances) (Arena et al., 2000; Bode, 1949; Carlson and Halijak, 1964; Nakagava and Sorimachi, 1992; Westerlund, 2002), control systems (Axtell and Bise, 1990; Dorčák, 1994; Podlubny, 1999b; Oustaloup, 1995), robotics (Marcos et al., 2008), signal processing (Tseng, 2007; Vinagre et al., 2003), chemical mixing (Oldham and Spanier, 1974), bioengineering (Magin, 2006), and so on. One of the very important areas of application is the chaos theory (West et al., 2002; Zaslavsky, 2005).

At this point we have to note that various mathematical definitions of chaos are known, but all of them express close characteristics of the dynamic systems that are concerned with supersensitivity or sensitive dependence on the initial conditions, which are characterized by Lyapunov instability as a main property of the chaotic oscillation. Roughly speaking, chaotic behaviour arises whenever the system trajectories are globally bounded and locally unstable (Andrievskii and Fradkov, 2003).

The fundamentals of a new mathematical apparatus for studying the chaotic phenomena and the theory of nonlinear oscillations were laid in 1960's and 1970's by A. Poincaré, B. Van der Pol, A. A. Andronov, N. M. Krylov, A. N. Kolmogorov, D. V. Anosov, Ya. G. Sinai, V. K. Mel'nikov, Yu. I. Neimark, L. P. Shil'nikov, G. M. Zaslavsky, and their collaborators. From that time on, the chaotic behavior has been discovered in numerous systems in mechanics, chemistry, physics, biology and medicine, electronic circuits, economics and so on (Andrievskii and Fradkov, 2004).

It is well known that chaos cannot occur in continuous nonlinear systems with the total order less than three (Silva, 1993). This assertion is based on the usual concepts of order, such as the number of states in a system or the total number of separate differentiations or integrations in the system. The model of chaotic system can be rearranged to three single differential equations, where the equations contain the non-integer (fractional) order derivatives. The total order of the system is the sum of each particular order instead of three. To put this fact into context, we can consider the fractional-order dynamical model of the system. Hartley et al. introduced the fractional-order Chua's system (Hartley et al., 1995). In the work (Arena et al., 1998), the fractional-order cellular neural network (CNN) was considered, the fractional Duffing's system was presented in the work (Gao and Yu, 2005), while other fractional-order chaotic systems were described in many other works (e.g., Ahmad, 2005; Deng et al., 2007; Guo, 2005; Li and Chen, 2004; Lu, 2005a,b; Nimmo and Evans, 1999, etc.). In all these cases chaos was exhibited in a system with total order less than three.

The term of "system order" should be mentioned and explained as well. The system order is not equal to the number of differential equations if we consider fractional differential equations. The system order is equal to the highest derivative of the fractional differential equation of the mathematical model. Arena et al. (1998) and Hartley et al. (1995) simply replaced the integer-order derivative by fractional order one. For numerical simulation they used an approximation method proposed by Charef et al. (1992). This approximation of fractional-order operators is in the form of rational polynomial of high order in the frequency domain. Such approximation may produce the so-called fake chaos. As what has been shown in the works (Tavazoei and Haeri, 2007, 2008), it is possible to calculate a minimal order of system, where chaos is still observed. In other cases it is a numerical error, which leads to the fake chaos. It is very important to take this into account.

The above considerations and conclusions in the aforementioned works by Arena et al. and Hartley et al. lead to several notes. We will discuss three of them. The first note is on fractional order of derivatives. We cannot simply replace the integer order with fractional order without any reasons. Some appropriate reasons are described in this book. It could be a fractional order in the capacitor model in the case of electrical chaotic circuits. The second note is on approximation methods used. If we use a high order approximation method then the total order of the system is not equal to the highest derivative of the fractional differential equation but is equal to the highest order of approximation polynomial. Moreover, the system could produce fake chaos through numerical errors in calculations. The third note is on system order. In most of the mentioned papers, the terms of system order, model order, number of initial conditions, number of state space variables and methods for rewriting the state-space representation to fractional differential equations are not clearly defined. In this book, we will discuss how to define them.

This book is also presented a collection of the Matlab functions created for numerical simulation of the described fractional-order chaotic systems. The function codes are downloadable from the website of MathWorks, Inc. and their utilization is described and commented in Appendix A of this book.

References

Ahmad W. M., 2005, Hyperchaos in fractional order nonlinear systems, *Chaos, Solitons and Fractals*, **26**, 1459–1465.

Andrievskii B. R. and Fradkov A. L., 2003, Control of chaos: Methods and applications. I. Methods, *Automation and Remote Control*, **64**, 673–713.

Andrievskii B. R. and Fradkov A. L., 2004, Control of chaos: Methods and applications. II. Applications, *Automation and Remote Control*, **65**, 505–533.

Arena P., Caponetto R., Fortuna L. and Porto D., 1998, Bifurcation and chaos in noninteger order cellular neural networks, *International Journal of Bifurcation and Chaos*, **8**, 1527–1539.

Arena P., Caponetto R., Fortuna L. and Porto D., 2000, *Nonlinear Noninteger Order Circuits and Systems – An Introduction*, World Scientific, Singapore.

Axtell M. and Bise E. M., 1990, Fractional calculus applications in control systems, *Proc. of the IEEE Nat. Aerospace and Electronics Conf.*, New York, 563–566.

Bode H. W., 1949, *Network Analysis and Feedback Amplifier Design*, Tung Hwa Book Company, Shanghai.

Cafagna D., 2007, Fractional Calculus: A mathematical tool from the past for present engineers, *IEEE Industrial Electronics Magazine*, **1**, 35–40.

Carlson G. E. and Halijak C. A., 1964, Approximation of fractional capacitors $(1/s)^{1/n}$ by a regular Newton process, *IEEE Trans. on Circuit Theory*, **11**, 210–213.

Charef A., Sun H. H., Tsao Y. Y. and Onaral B., 1992, Fractal system as represented by singularity function, *IEEE Transactions on Automatic Control*, **37**, 1465–1470.

Deng W., Li C. and Lu J., 2007, Stability analysis of linear fractional differential system with multiple time delays, *Nonlinear Dynamics*, **48**, 409–416.

Dorčák Ľ., 1994, *Numerical Models for the Simulation of the Fractional-Order Control Systems, UEF-04-94*, The Academy of Sciences, Inst. of Experimental Physic, Košice, Slovakia.

Gao X. and Yu J., 2005, Chaos in the fractional order periodically forced complex Duffing's oscillators, *Chaos, Solitons and Fractals*, **24**, 1097–1104.

Guo L. J., 2005, Chaotic dynamics and synchronization of fractional-order Genesio-Tesi systems, *Chinese Physics*, **14**, 1517–1521.

Hartley T. T., Lorenzo C. F. and Qammer H. K., 1995, Chaos on a fractional Chua's system, *IEEE Trans. Circ. Syst. Fund. Theor. Appl.*, **42**, 485–490.

Li C. and Chen G., 2004, Chaos and hyperchaos in the fractional-order Rossler equations, *Physica A*, **341**, 55–61.

Lu J. G., 2005a, Chaotic dynamics and synchronization of fractional-order Arneodo's systems, *Chaos, Solitons and Fractals*, **26**, 1125–1133.

Lu J. G., 2005b, Chaotic dynamics and synchronization of fractional-order Chua's circuits with a piecewise-linear nonlinearity, *Int. Journal of Modern Physics B*, **19**, 3249–3259.

Magin R. L., 2006, *Fractional Calculus in Bioengineering*, Begell House Publishers, Redding.

da Graca Marcos, M., Duarte F. B. M. and Machado J. A. T., 2008, Fractional dynamics in the trajectory control of redundant manipulators, *Communications in Nonlinear Science and Numerical Simulations*, **13**, 1836–1844.

Miller K. S. and Ross B., 1993, *An Introduction to the Fractional Calculus and Fractional Differential Equations*, John Wiley & Sons. Inc., New York.

Nakagava M. and Sorimachi K., 1992, Basic characteristics of a fractance device, *IEICE Trans. fundamentals*, **E75-A**, 1814–1818.

Nimmo S. and Evans A. K., 1999, The effects of continuously varying the fractional differential order of chaotic nonlinear systems, *Chaos, Solitons and Fractals*, **10**, 1111–1118.

Oldham K. B. and Spanier J., 1974, *The Fractional Calculus*, Academic Press, New York.

Oustaloup A., 1995, *La Derivation Non Entiere: Theorie, Synthese et Applications*, Hermes, Paris.

Parada F. J. V., Tapia J. A. O. and Ramirez J. A., 2007, Effective medium equations for fractional Fick's law in porous media, *Physica A*, **373**, 339–353.

Podlubny I., 1999a, *Fractional Differential Equations*, Academic Press, San Diego.

Podlubny I., 1999b, Fractional-order systems and $PI^{\lambda}D^{\mu}$-controllers, *IEEE Transactions on Automatic Control*, **44**, 208–213.

Silva C. P., 1993, Shil'nikov's theorem - A tutorial, *IEEE Transactions on Circuits and Systems-I: Fundamental Theory and Applications*, **40**, 675–682.

Tavazoei M. S. and Haeri M., 2007, Unreliability of frequency-domain approximation in recognising chaos in fractional-order systems, *IET Signal Proc.*, **1**, 171–181.

Tavazoei M. S. and Haeri M., 2008, Limitations of frequency domain approximation for detecting chaos in fractional order systems, *Nonlinear Analysis*, **69**, 1299–1320.

Torvik P. J. and Bagley R. L., 1984, On the appearance of the fractional derivative in the behavior of real materials, *Transactions of the ASME*, **51**, 294–298.

Tseng C. C., 2007, Design of FIR and IIR fractional order Simpson digital integrators, *Signal Processing*, **87**, 1045–1057.

Vinagre B. M., Chen Y. Q. and Petráš I., 2003, Two direct Tustin discretization methods for fractional-order differentiator/integrator, *J. Franklin Inst.*, **340**, 349–362.

Wang J. C., 1987, Realizations of generalized Warburg impedance with RC ladder networks and transmission lines, *Journal of The Electrochemical Society*, **134**, 1915–1920.

West B. J., Bologna M. and Grigolini P., 2002, *Physics of Fractal Operators*, Springer, New York.

Westerlund S., 2002, *Dead Matter Has Memory!*, Causal Consulting, Kalmar, Sweden.

Zaslavsky G. M., 2005, *Hamiltonian Chaos and Fractional Dynamics*, Oxford University Press, Oxford.

Chapter 2
Fractional Calculus

2.1 Special Functions

Here, we should mention the most important function used in fractional calculus — Euler's *Gamma* function, which is defined as

$$\Gamma(n) = \int_0^\infty t^{n-1}\, e^{-t}\, dt. \tag{2.1}$$

This function is generalization of a factorial in the following form:

$$\Gamma(n) = (n-1)! \tag{2.2}$$

Another function, which plays a very important role in the fractional calculus, was in fact introduced by Humbert and Agarwal in 1953. It is a two-parameter function of the *Mittag-Leffler* type defined as (Gorenflo et al., 1998):

$$E_{\alpha,\beta}(z) = \sum_{k=0}^{\infty} \frac{z^k}{\Gamma(\alpha k+\beta)}, \qquad \alpha > 0, \quad \beta > 0. \tag{2.3}$$

There are some relationships (e.g. in Djrbashian, 1993; Podlubny, 1999):

$$\begin{aligned}
&E_{1,1}(z) = e^z, \qquad E_{1,2}(z) = \frac{e^z-1}{z},\\
&E_{2,1}(z) = \cosh(\sqrt{z}), \qquad E_{2,1}(-z^2) = \cos(z),\\
&E_{1/2,1}(\sqrt{z}) = \frac{2}{\sqrt{\pi}} e^{-z}\mathrm{erfc}(-\sqrt{z}).
\end{aligned}$$

For $\beta = 1$ we obtain the Mittag-Leffler function in one parameter (Mittag-Leffler, 1903):

$$E_{\alpha,1}(z) = \sum_{k=0}^{\infty} \frac{z^k}{\Gamma(\alpha k+1)} \equiv E_\alpha(z). \tag{2.4}$$

For the numerical evaluation of the Mittag-Leffler function with the default accuracy set to 10^{-6} the Matlab routine written by Podlubny and Kacenak can be used (Podlubny and Kacenak, 2005). Figures 2.1, 2.2 are plotted some well-known functions ($\cos(z)$ and e^z).

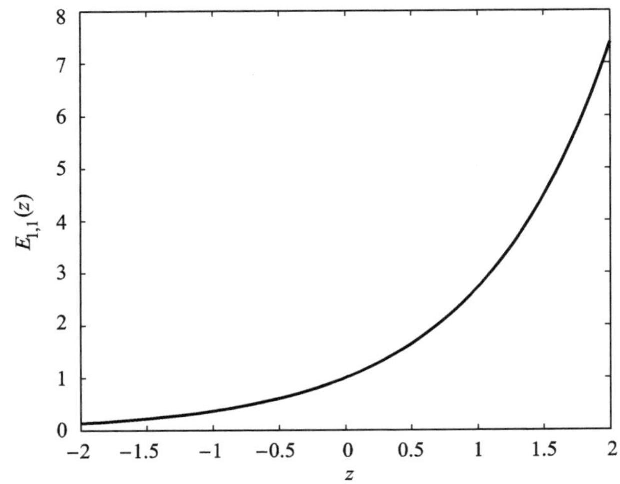

Fig. 2.1 Mittag-Leffler function for parameters $E_{1,1}(z)$, where $-2 < z < 2$

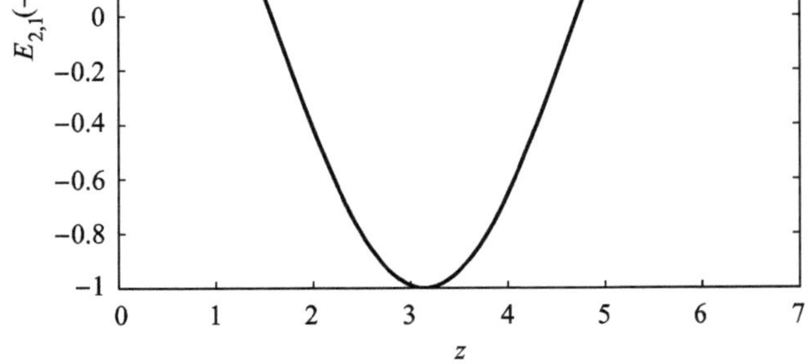

Fig. 2.2 Mittag-Leffler function for parameters $E_{2,1}(-z^2)$, where $0 < z < 2\pi$

2.2 Definitions of Fractional Derivatives and Integrals

Fractional calculus is a generalization of integration and differentiation to noninteger-order fundamental operator ${}_aD_t^\alpha$, where a and t are the bounds of the operation and $\alpha \in$ R. The continuous integro-differential operator is defined as

$$
{}_aD_t^\alpha = \begin{cases} \dfrac{d^\alpha}{dt^\alpha}, & \alpha > 0, \\ 1, & \alpha = 0, \\ \displaystyle\int_a^t (d\tau)^\alpha, & \alpha < 0. \end{cases}
$$

The three most frequently used definitions for the general fractional differintegral are: the Grünwald-Letnikov (GL) definition, the Riemann-Liouville (RL) and the Caputo definitions (Miller and Ross, 1993; Oldham and Spanier, 1974; Podlubny, 1999). Other definitions are connected with well-known names as, for instance, Weyl, Fourier, Cauchy, Abel, Nishimoto, etc.

In this book we will consider mainly the GL, the RL, and the Caputo's definitions. This consideration is based on the fact that for a wide class of functions, the three best known definitions — GL, RL, and Caputo — are equivalent under some conditions (Podlubny, 1999).

2.3 Grünwald-Letnikov Fractional Integrals and Derivatives

Let us consider the continuous function $f(t)$. Its first derivative can be expressed as

$$
\frac{d}{dt} f(t) \equiv f'(t) = \lim_{h \to 0} \frac{f(t) - f(t-h)}{h} \tag{2.5}
$$

By using Eq. (2.5) twice, we obtain a second derivative of the function $f(t)$ in the form

$$
\begin{aligned}
\frac{d^2}{dt^2} f(t) \equiv f''(t) &= \lim_{h \to 0} \frac{f'(t) - f'(t-h)}{h} \\
&= \lim_{h \to 0} \frac{1}{h} \left\{ \frac{f(t) - f(t-h)}{h} - \frac{f(t-h) - f(t-2h)}{h} \right\} \\
&= \lim_{h \to 0} \frac{f(t) - 2f(t-h) + f(t-2h)}{h^2}.
\end{aligned} \tag{2.6}
$$

With (2.5) and (2.6) we can get a third derivative of the function $f(t)$ as

$$
\frac{d^3}{dt^3} \equiv f'''(t) = \lim_{h \to 0} \frac{f(t) - 3f(t-h) + 3f(t-2h) - f(t-3h)}{h^3}. \tag{2.7}
$$

According to this rule we can write a general formula for n-derivative of the function $f(t)$ by t for $n \in \mathrm{N}$, $j > n$ in the following form:

$$\frac{d^n}{dt^n} f(t) \equiv f^{(n)}(t) = \lim_{h \to 0} \frac{1}{h^n} \sum_{j=0}^{n} (-1)^j \binom{n}{j} f(t - jh). \tag{2.8}$$

The relation (2.8) expresses a linear combination of function values $f(t)$ in variable t. Binomial coefficients with alternating signs for positive value of n are defined as

$$\binom{n}{j} = \frac{n(n-1)(n-2)\cdots(n-j+1)}{j!} = \frac{n!}{j!(n-j)!}. \tag{2.9}$$

In the case of negative value of n we have

$$\binom{-n}{j} = \frac{-n(-n-1)(-n-2)\cdots(-n-j+1)}{j!} = (-1)^j \begin{bmatrix} n \\ j \end{bmatrix}, \tag{2.10}$$

where $\begin{bmatrix} n \\ j \end{bmatrix}$ is defined as

$$\begin{bmatrix} n \\ j \end{bmatrix} = \frac{n(n+1)\cdots(n+j-1)}{j!} \tag{2.11}$$

If we substitute $-n$ in (2.8) for n, we can write

$$\frac{d^{-n}}{dt^{-n}} f(t) \equiv f^{(-n)}(t) = \lim_{h \to 0} \frac{1}{h^n} \sum_{j=0}^{n} \begin{bmatrix} n \\ j \end{bmatrix} f(t - jh), \tag{2.12}$$

where n is a positive integer number.

According to Eqs. (2.5)—(2.8) we can write the fractional-order derivative definition of order α, $(\alpha \in \mathrm{R})$ by t, which has the form

$$D_t^{\alpha} f(t) = \lim_{h \to 0} \frac{1}{h^{\alpha}} \sum_{j=0}^{\infty} (-1)^j \binom{\alpha}{j} f(t - jh). \tag{2.13}$$

For binomial coefficients calculation we can use the relation between Euler's *Gamma* function and factorial, defined as

$$\binom{\alpha}{j} = \frac{\alpha!}{j!(\alpha - j)!} = \frac{\Gamma(\alpha+1)}{\Gamma(j+1)\Gamma(\alpha - j + 1)} \tag{2.14}$$

for $\binom{\alpha}{0} = 1$.

If we consider $n = \frac{t-a}{h}$, where a is a real constant, which expresses a limit value, we can write

$${}_aD_t^{\alpha} f(t) = \lim_{h \to 0} \frac{1}{h^{\alpha}} \sum_{j=0}^{[\frac{t-a}{h}]} (-1)^j \binom{\alpha}{j} f(t - jh), \tag{2.15}$$

where $[x]$ means the integer part of x, a and t are the bounds of operation for ${}_aD_t^\alpha f(t)$.

2.4 Riemann-Liouville Fractional Integrals and Derivatives

For expression of the Riemann-Liouville definition, we will consider the Riemann - Liouville n - fold integral defined as

$$\underbrace{\int_a^t \int_a^{t_n} \int_a^{t_{n-1}} \cdots \int_a^{t_3} \int_a^{t_2}}_{n-\text{fold}} f(t_1) dt_1 dt_2 \cdots dt_{n-1} dt_n = \frac{1}{\Gamma(n)} \int_a^t \frac{f(\tau)}{(t-\tau)^{1-n}} d\tau \quad (2.16)$$

for $n \in \mathrm{N}$, $n > 0$.

Fractional integral of order α for the function $f(t)$ can be expressed from Eq. (2.16) as follows:

$${}_aI_t^\alpha f(t) \equiv {}_aD_t^{-\alpha} f(t) = \frac{1}{\Gamma(-\alpha)} \int_a^t \frac{f(\tau)}{(t-\tau)^{\alpha+1}} d\tau \quad (2.17)$$

for $\alpha, a \in \mathrm{R}$, $\alpha < 0$.

From relation (2.17) we can write formula for the Riemann-Liouville definition of fractional derivative of the order α in the following form

$${}_aD_t^\alpha f(t) = \frac{1}{\Gamma(n-\alpha)} \frac{d^n}{dt^n} \int_a^t \frac{f(\tau)}{(t-\tau)^{\alpha-n+1}} d\tau \quad (2.18)$$

for $(n-1 < \alpha < n)$, where a and t are the limits of operation ${}_aD_t^\alpha f(t)$.

For the case of $0 < \alpha < 1$ and $f(t)$ being a causal function of t, that is, $f(t) = 0$ for $t < 0$, the fractional integral is defined as:

$${}_0D_t^{-\alpha} f(t) = \frac{1}{\Gamma(\alpha)} \int_0^t \frac{f(\tau)}{(t-\tau)^{1-\alpha}} d\tau, \text{ for } 0 < \alpha < 1,\ t > 0 \quad (2.19)$$

and the expression for the fractional order derivative is:

$${}_0D_t^\alpha f(t) = \frac{1}{\Gamma(n-\alpha)} \frac{d^n}{dt^n} \int_0^t \frac{f(\tau)}{(t-\tau)^{\alpha-n+1}} d\tau, \quad (2.20)$$

where $\Gamma(\cdot)$ is Euler's *Gamma* function (Oldham and Spanier, 1974).

2.5 Caputo Fractional Derivatives

The Caputo definition of fractional derivatives can be written as (Caputo, 1967; Podlubny, 1999):

$$ {}_aD_t^{\alpha} f(t) = \frac{1}{\Gamma(n-\alpha)} \int_a^t \frac{f^{(n)}(\tau)}{(t-\tau)^{\alpha-n+1}} d\tau, \quad \text{for } n-1 < \alpha < n. \tag{2.21} $$

As mentioned above, under the homogenous initial conditions the Riemann-Liouville and the Caputo derivatives are equivalent. Let us denote the Riemann-Liouville fractional derivative by ${}_a^{RL}D_t^{\alpha} t(t)$ and the Caputo defnition by ${}_a^{C}D_t^{\alpha} f(t)$, then the relationship between them is:

$$ {}_a^{RL}D_t^{\alpha} t(t) = {}_a^{C}D_t^{\alpha} f(t) + \sum_{k=0}^{n-1} \frac{(t-a)^{k-\alpha}}{\Gamma(k-\alpha+1)} f^{(k)}(a) $$

for $f^{(k)}(a) = 0$, $(k = 0, 1, \ldots, n-1)$.

The initial conditions for the fractional-order differential equations with the Caputo derivatives are in the same form as for the integer-order differential equations. It is an advantage because applied problems require definitions of fractional derivatives, where there are clear interpretations of initial conditions, which contain $f(a)$, $f^{'}(a)$, $f^{''}(a)$, etc.

2.6 Laplace Transform Method

The Laplace transform method is a very frequently used tool for solving engineering problems. In this section we will recall some basic facts about the Laplace transform method for integer order and then we will show this method for fractional order as well.

2.6.1 Basic Facts about the Laplace Transform

The function $F(s)$ of the complex variable s defined by

$$ F(s) = L\{f(t); s\} = \int_0^{\infty} e^{-st} f(t) dt \tag{2.22} $$

is called the Laplace transform of function $f(t)$, which is called the original. For the existence of integral (2.22) the function $f(t)$ must be of exponential order α, which means that there exist positive constants M and T so that

$$e^{-\alpha t}|f(t)| \leq M, \quad \text{for all} \quad t > T.$$

In other words, the function $f(t)$ must not grow faster than a certain exponential function when $t \to \infty$.

We will denote the Laplace transforms by the uppercase letters and the originals by the lowercase letters.

The original $f(t)$ can be restored from the Laplace transform $F(s)$ with the help of the inverse Laplace transform

$$f(t) = L^{-1}\{F(s);t\} = \int_{c-j\infty}^{c+j\infty} e^{st}F(s)ds, \quad c = Re(s) > c_0, \tag{2.23}$$

where c_0 lies in the right half plane of the absolute convergence of the Laplace integral (2.22).

The direct evaluation of the inverse Laplace transform using the formula (2.23) is often complicated, however, sometimes it gives useful information on the behavior of the unknown original $f(t)$ we look for.

The Laplace transform of the convolution

$$f(t) * g(t) = \int_0^t f(t-\tau)g(\tau)d\tau = \int_0^t f(\tau)g(t-\tau)d\tau \tag{2.24}$$

of the two functions $f(t)$ and $g(t)$, which are equal to zero for $t < 0$, is equal to the product of the Laplace transform of those functions:

$$L\{f(t) * g(t);s\} = F(s)G(s) \tag{2.25}$$

under the assumption that both $F(s)$ and $G(s)$ exist. We will use the property (2.25) for the evaluation of the Laplace transform of the Riemann-Liouville fractional integral.

Another useful property we need is the formula for the Laplace transform of the derivative of an integer order n of the function $f(t)$:

$$L\{f^n(t);s\} = s^nF(s) - \sum_{k=0}^{n-1} s^{n-k-1}f^{(k)}(0) = s^nF(s) - \sum_{k=0}^{n-1} s^k f^{(n-k-1)}(0), \tag{2.26}$$

which can be obtained from the definition (2.22) by integrating by parts under the assumption that the corresponding integrals exist.

In the following sections relating the Laplace transforms of the fractional derivatives we consider the lower bound $a = 0$.

2.6.2 Laplace Transform of Fractional Integrals

We will start with the Laplace transform of the Riemann-Liouville fractional integral of order $p > 0$ defined by (2.19), which we can write as a convolution of the functions $g(t) = t^{p-1}$ and $f(t)$:

$$ {}_0D_t^{-p} f(t) = {}_0D_t^{-p} f(t) = \frac{1}{\Gamma(p)} \int_0^t (t-\tau)^{p-1} f(\tau) d\tau = t^{p-1} * f(t). \tag{2.27} $$

The Laplace transform of the function t^{p-1} is (Podlubny, 1999):

$$ G(s) = L\{t^{p-1}; s\} = \Gamma(p) s^{-p}. \tag{2.28} $$

Therefore, using the formula for the Laplace transform of the convolution (2.25) we obtain the Laplace transform of the Riemann-Liouville and the Grünwald-Letnikov fractional integrals:

$$ L\{{}_0D_t^{-p} f(t); s\} = L\{{}_0D_t^{-p} f(t); s\} = s^{-p} F(s). \tag{2.29} $$

2.6.3 Laplace Transform of Fractional Derivatives

Now let us turn to the evaluation of the Laplace transform of the Riemann-Liouville fractional derivative, which we, for this purpose, write in the form:

$$ {}_0D_t^p f(t) = g^{(n)}(t), \tag{2.30} $$

$$ g(t) = {}_0D_t^{-(n-p)} f(t) \frac{1}{\Gamma(k-p)} \int_0^t (t-\tau)^{n-p-1} f(\tau) d\tau, \tag{2.31} $$

$$ n - 1 \leq p < n. $$

The use of the formula for the Laplace transform of an integer-order derivative (2.26) leads to

$$ L\{{}_0D_t^p f(t); s\} = s^n G(s) - \sum_{k=0}^{n-1} s^k g^{(n-k-1)}(0). \tag{2.32} $$

The Laplace transform of the function $g(t)$ is evaluated by (2.29):

$$ G(s) = s^{-(n-p)} F(s). \tag{2.33} $$

Additionally, from the definition of the Riemann-Liouville fractional derivative (2.20) it follows that

$$g^{(n-k-1)}(t) = \frac{d^{n-k-1}}{dt^{n-k-1}}\, {}_0D_t^{-(n-p)} f(t) = {}_0D_t^{p-k-1} f(t). \tag{2.34}$$

Substituting (2.33) and (2.34) into (2.32) we obtain the following final expression for the Laplace transform of the Riemann-Liouville fractional derivative of order $p > 0$:

$$L\{{}_0D_t^p f(t); s\} = s^p F(s) - \sum_{k=0}^{n-1} s^k \left[{}_0D_t^{p-k-1} f(t)\right]_{t=0}. \tag{2.35}$$

$$n-1 \leq p < n.$$

The formula for the Laplace transform of the Caputo fractional derivative (2.18) has the form (Podlubny, 1999):

$$\int_0^\infty e^{-st}\, {}_0D_t^\alpha f(t)\, dt = s^\alpha F(s) - \sum_{k=0}^{n-1} s^{\alpha-k-1} f^{(m)}(0), \quad n-1 \leq \alpha < n. \tag{2.36}$$

For zero initial conditions, the Laplace transform of fractional derivatives of order r (Grünwald-Letnikov, Riemann-Liouville, and Caputo's) reduces to:

$$L\{{}_0D_t^r f(t)\} = s^r F(s). \tag{2.37}$$

Morever, the Laplace transform of the Riemann-Liouville fractional derivative is well known. However, its practical applicability is limited by the absense of the physical interpretation of the limit values of fractional derivatives at the lower bound $t = 0$. So far, such interpretation was partially solved only in the work done by Heymans and Podlubny, in 2006.

2.7 Fourier Transform Method

2.7.1 Basic Facts about the Fourier Transform

The exponential Fourier transform of a continuous function $h(t)$ absolutely integrable in $(-\infty, \infty)$ is defined by

$$F_e\{h(t); \omega\} = \int_{-\infty}^{\infty} e^{j\omega t} h(t) dt, \tag{2.38}$$

and the original $h(t)$ can be restored from its Fourier transform $H_e(t)$ with the help of the inverse Fourier transform:

$$h(t) = \frac{1}{2\pi} \int_{-\infty}^{\infty} H_e(\omega) e^{-j\omega t} d\omega. \tag{2.39}$$

As mentioned above, we will denote originals by the lowercase letters, and their transforms by the uppercase letters.

The Fourier transform of the convolution

$$h(t)*g(t) = \int_{-\infty}^{\infty} h(t-\tau)g(\tau)d\tau = \int_{-\infty}^{\infty} h(\tau)g(t-\tau)d\tau \tag{2.40}$$

of the two functions $h(t)$ and $g(t)$, which are defined in $(-\infty, \infty)$, is equal to the product of their Fourier transforms:

$$F_e\{h(t)*g(t);\, \omega\} = H_e(\omega)\, G_e(\omega) \tag{2.41}$$

under the assumption that both $H_e(\omega)$ and $G_e(\omega)$ exist. We will use the property (2.41) for the evaluation of Fourier transforms of the Riemann-Liouville fractional integral and Fourier transforms of fractional derivatives.

Another useful property of the Fourier transform, which is frequently used in solving applied problems, is the Fourier transform of derivatives of $h(t)$. Namely, if $h(t)$, $h'(t)$, ..., $h^{(n-1)}(t)$ vanish for $t \to \pm\infty$, then the Fourier transform of the n-th derivative of $h(t)$ is

$$F_e\{h^{(n)}(t);\, \omega\} = (-j\omega)^n H_e(\omega) \tag{2.42}$$

The Fourier transform is a powerful tool for frequency domain analysis of linear dynamical systems.

2.7.2 Fourier Transform of Fractional Integrals

First we will evaluate the Fourier transform of the Riemann-Liouville fractional integral with the lower terminal $a = -\infty$, i.e. of

$${}_{-\infty}D_t^{-\alpha} g(t) = \frac{1}{\Gamma(\alpha)} \int_{-\infty}^{t} (t-\tau)^{\alpha-1} g(\tau)d\tau, \tag{2.43}$$

where we assume $0 < \alpha < 1$.

Let us start with the Laplace transform of the function

$$h(t) = \frac{t^{\alpha-1}}{\Gamma(\alpha)}$$

(see formula (2.28)), which can be written as

$$\frac{1}{\Gamma(\alpha)} \int_{0}^{\infty} t^{\alpha-1} e^{-st} dt = s^{-\alpha}. \tag{2.44}$$

Let us take $s = -j\omega$, where ω is real. It follows from the Dirichlet theorem that in such a case the integral (2.44) converges if $0 < \alpha < 1$. Therefore, we immediately obtain the Fourier transform of the function

$$h_+(t) = \begin{cases} \dfrac{t^{\alpha-1}}{\Gamma(\alpha)}, & t > 0, \\ 0, & t \leq 0, \end{cases}$$

in the form

$$F_e\{h_+(t);\, \omega\} = (-j\omega)^{-\alpha}. \tag{2.45}$$

Now we can find the Fourier transform of the Riemann-Liouville fractional integral (2.43), which can be written as a convolution (2.40) of the functions $h_+(t)$ and $g(t)$:

$${}_{-\infty}D_t^{-\alpha} f(t) = h_+(t) * g(t). \tag{2.46}$$

Using the rule (2.41) we obtain:

$$F_e\left\{{}_{-\infty}D_t^{-\alpha} g(t);\, \omega\right\} = (j\omega)^{-\alpha} G(\omega), \tag{2.47}$$

where $G(\omega)$ is the Fourier transform of the function $g(t)$.

The formula (2.47) gives also the Fourier transform of the Grünwald-Letnikov fractional integral ${}_{-\infty}D_t^{-\alpha} g(t)$, because in the considered case it coincides with the Riemann-Liouville fractional integral.

2.7.3 Fourier Transform of Fractional Derivatives

Let us now evaluate the Fourier transform of fractional derivatives.

Considering the lower terminal $a = -\infty$ and requiring the resonable behavior of $g(t)$ and its derivatives for $t \to -\infty$ we can perform integration by parts and write the Riemann-Liouville in the form:

$${}_{-\infty}D_t^{\alpha} g(t) = \frac{1}{\Gamma(n-\alpha)} \int_{-\infty}^{t} \frac{g^{(n)}(\tau)\, d\tau}{(t-\tau)^{\alpha+1-n}} = {}_{-\infty}D_t^{\alpha-n} g^{(n)}(t), \tag{2.48}$$

$$n-1 < \alpha < n.$$

The Fourier transform of (2.48) with the use of the Fourier transform of the Riemann-Liouville fractional integral (2.47) and then the Fourier transform of an integer-order derivative (2.42) gives the following formula for the exponential Fourier transform of the Riemann-Liouville with the lower terminal $a = -\infty$:

$$\begin{aligned} F_e\{D^{\alpha} g(t);\, \omega\} &= (-j\omega)^{\alpha-n} F_e\left\{g^{(n)}(t);\, \omega\right\} \\ &= (-j\omega)^{\alpha-n}(-j\omega)^n G(\omega) \\ &= (-j\omega)^{\alpha} G(\omega). \end{aligned} \tag{2.49}$$

2.8 Some Properties of Fractional Derivatives and Integrals

The main properties of fractional derivatives/integrals are as follows (Oldham and Spanier, 1974):

1. If $f(t)$ is an analytical function of t, then its fractional derivative ${}_0D_t^{\alpha} f(t)$ is an analytical function of t, α.
2. For $\alpha = n$, where n is integer, the operation ${}_0D_t^{\alpha} f(t)$ gives the same result as classical differentiation of integer order n.
3. For $\alpha = 0$ the operation ${}_0D_t^{\alpha} f(t)$ is the identity operator:
$$ {}_0D_t^{0} f(t) = f(t). $$
4. Fractional differentiation and fractional integration are linear operations:
$$ {}_aD_t^{\alpha}\left(\lambda f(t) + \mu g(t)\right) = \lambda\, {}_aD_t^{\alpha} f(t) + \mu\, {}_aD_t^{\alpha} g(t). \tag{2.50} $$
5. The additive index law (semigroup property)
$$ {}_0D_t^{\alpha}\, {}_0D_t^{\beta} f(t) = {}_0D_t^{\beta}\, {}_0D_t^{\alpha} f(t) = {}_0D_t^{\alpha+\beta} f(t) $$
holds under some reasonable constraints on the function $f(t)$.
The fractional-order derivative commutes with integer-order derivative
$$ \frac{d^n}{dt^n}\left({}_aD_t^{r} f(t)\right) = {}_aD_t^{r}\left(\frac{d^n f(t)}{dt^n}\right) = {}_aD_t^{r+n} f(t), \tag{2.51} $$
under the condition $t = a$ we have $f^{(k)}(a) = 0$, $(k = 0, 1, 2, \ldots, n-1)$. The relationship above says the operators $\frac{d^n}{dt^n}$ and ${}_aD_t^{r}$ commute.
6. The Leibniz's rule for fractional differentiation is given as:
$$ {}_aD_t^{r}\left(\phi(t) f(t)\right) = \sum_{k=0}^{\infty} \binom{r}{k} \phi^{(k)}(t)\, {}_aD_t^{r-k} f(t), \tag{2.52} $$
if $\phi(t)$ and $f(t)$ and all their derivatives are continuous in the interval $[a,t]$.
7. Geometric and physical interpretation of fractional integration and fractional differentiation was clearly explained in Podlubny's work (Podlubny, 2002).

Some other important properties of fractional derivatives and integrals as for example translation, chain rule, behavior and dependence on limit and so on can be found in several other works (e.g., Miller and Ross, 1993; Oldham and Spanier, 1974; Oustaloup, 1995; Podlubny, 1999, etc.).

2.9 Numerical Methods for Calculation of Fractional Derivatives and Integrals

For numerical calculation of fractional-order derivatives we can use the relation (2.53) derived from the GL definition (2.15). This approach is based on the fact that for a wide class of functions, three definitions — GL (2.15), RL (2.18), and Caputo's (2.21) — are equivalent. The relation to the explicit numerical approximation of q-th derivative at the points kh, $(k = 1, 2, \ldots)$ has the following form (Dorčák, 1994; Podlubny, 1999; Vinagre et al., 2003):

$$ {}_{(k-L_m/h)}D^q_{t_k} f(t) \approx h^{-q} \sum_{j=0}^{k} (-1)^j \binom{q}{j} f(t_{k-j}), \tag{2.53} $$

where L_m is the "memory length", $t_k = kh$, h is the time step of calculation and $(-1)^j \binom{q}{j}$ are binomial coefficients $c_j^{(q)}$ $(j = 0, 1, \ldots)$. For their calculation we can use the following expression (Dorčák, 1994):

$$ c_0^{(q)} = 1, \qquad c_j^{(q)} = \left(1 - \frac{1+q}{j}\right) c_{j-1}^{(q)}. \tag{2.54} $$

Then, general numerical solution of the fractional differential equation

$$ {}_aD_t^q y(t) = f(y(t), t), $$

can be expressed as

$$ y(t_k) = f\left(y(t_k), t_k\right) h^q - \sum_{j=v}^{k} c_j^{(q)} y(t_{k-j}). \tag{2.55} $$

For the *memory term* expressed by the sum, a "short memory" principle can be used. Then the lower index of the sums in relations (2.55) will be $v = 1$ for $k < (L_m/h)$ and $v = k - (L_m/h)$ for $k > (L_m/h)$, or without using the "short memory" principle, we put $v = 1$ for all k.

Obviously, for this simplification we pay a penalty in the form of some inaccuracy. If $f(t) \le M$, we can easily establish the following estimate for determining the memory length L_m, providing the required accuracy ε:

$$ L_m \ge \left(\frac{M}{\varepsilon|\Gamma(1-q)|}\right)^{1/q}. \tag{2.56} $$

An evaluation of the short memory effect and convergence relation of the error between short and long memory were clearly described and also proved in (Podlubny, 1999).

Described numerical method is the so-called Power Series Expansion (PSE) of a generating function. It is important to note that PSE leads to approximation in the form of polynomials, that is, the discretized fractional operator is in the form of FIR filter, which has only zeros.

The resulting discrete transfer function, approximating fractional-order operators, can be expressed in z-domain as follows:

$$ {}_0D_{kT}^{\pm r}G(z) = \frac{Y(z)}{F(z)} = \left(\frac{1}{T}\right)^{\pm r} \mathrm{PSE}\left\{\left(1-z^{-1}\right)^{\pm r}\right\}_n \approx T^{\mp r}R_n(z^{-1}), \tag{2.57}$$

where T is the sample period, PSE$\{u\}$ denotes the function resulting from applying the power series expansion to the function u, $Y(z)$ is the Z transform of the output sequence $y(kT)$, $F(z)$ is the Z transform of the input sequence $f(kT)$, n is the order of the approximation, and R is polynomial of degree n, respectively, in the variable z^{-1}, $z = \exp(sT)$, and $k = 1,2,\ldots$. Matlab routine `dfod1()` of this method can be downloaded from MathWorks, Inc. web site (Petráš, 2003b).

Other approach can be realized by Continued Fraction Expansion (CFE) of the generating function and then the approximated fractional operator is in the form of IIR filter, which has poles and zeros (Chen et al., 2006).

Taking into account that our aim is to obtain equivalents to the fractional integrodifferential operators in the Laplace domain, $s^{\pm r}$, the result of such approximation for an irrational function, $G(s)$, can be expressed in the form:

$$\begin{aligned} G(s) &\simeq a_0(s) + \cfrac{b_1(s)}{a_1(s) + \cfrac{b_2(s)}{a_2(s) + \cfrac{b_3(s)}{a_3(s)+\cdots}}} \\ &= a_0(s) + \frac{b_1(s)}{a_1(s)+}\,\frac{b_2(s)}{a_2(s)+}\,\frac{b_3(s)}{a_3(s)+}\,\cdots, \end{aligned} \tag{2.58}$$

where $a_i(s)$ and $b_i(s)$ are rational functions of the variable s, or are constants. The application of the method yields a rational function, which is an approximation of the irrational function $G(s)$.

In other words, for evaluation purposes, the rational approximations obtained by CFE frequently converge much more rapidly than the PSE and have a wider domain of convergence in the complex plane. On the other hand, the approximation by PSE and the short memory principle is convenient for the dynamical properties consideration.

For interpolation purposes, rational functions are sometimes superior to polynomials. This is, roughly speaking, due to their ability to model functions with poles. These techniques are based on the approximations of an irrational function, $G(s)$, by a rational function defined by the quotient of two polynomials in the variable s in frequency s-domain:

$$\begin{aligned} G(s) \simeq R_{i(i+1)\cdots(i+m)} &= \frac{P_\mu(s)}{Q_\nu(s)} = \frac{p_0 + p_1 s + \cdots + p_\mu s^\mu}{q_0 + q_1 s + \cdots + q_\nu s^\nu}, \\ m+1 &= \mu + \nu + 1, \end{aligned} \tag{2.59}$$

passing through the points $(s_i, G(s_i)), \ldots, (s_{i+m}, G(s_{i+m}))$.

The resulting discrete transfer function, approximating fractional-order operators, can be expressed as (Vinagre et al., 2003):

$$
\begin{aligned}
{}_0D_{kT}^{\pm r} G(z) &= \frac{Y(z)}{F(z)} = \left(\frac{2}{T}\right)^{\pm r} \mathrm{CFE}\left\{\left(\frac{1-z^{-1}}{1+z^{-1}}\right)^{\pm r}\right\}_{p,n} \\
&\approx \left(\frac{2}{T}\right)^{\pm r} \frac{P_p(z^{-1})}{Q_n(z^{-1})}
\end{aligned}
\tag{2.60}
$$

where T is the sample period, $\mathrm{CFE}\{u\}$ denotes the function resulting from applying the continued fraction expansion to the function u, $Y(z)$ is the Z transform of the output sequence $y(kT)$, $F(z)$ is the Z transform of the input sequence $f(kT)$, p and n are the orders of the approximation, and P and Q are polynomials of degrees p and n, respectively, in the variable z^{-1}, and $k = 1, 2, \ldots$. Matlab routine `dfod2()` can be downloaded from MathWorks, Inc. web site (Petráš, 2003a).

In general, the discretization of fractional-order differentiator/integrator $s^{\pm r}$ $(r \in \mathrm{R})$ can be expressed by the *generating function* $s \approx \omega(z^{-1})$. This generating function and its expansion determine both the form of the approximation and the coefficients (Lubich, 1986).

As a generating function $\omega(z^{-1})$ can be used in general in the following formula (Barbosa et al., 2006):

$$
\omega(z^{-1}) = \left(\frac{1}{\beta T} \frac{1-z^{-1}}{\gamma + (1-\gamma)z^{-1}}\right), \tag{2.61}
$$

where β and γ denote the gain and phase tuning parameters, respectively. For example, when $\beta = 1$ and $\gamma = \{0, 1/2, 7/8, 1, 3/2\}$, the generating function (2.61) becomes the forward Euler, the Tustin, the Al-Alaoui, the backward Euler, the implicit Adams rules, respectively. In this sense the generating formula can be tuned precisely. The expansion of the generating functions can be done by PSE or CFE.

Taking into account that our aim is to obtain discrete equivalents to the fractional integrodifferential operators in the Laplace domain, $s^{\pm r}$, the following considerations have to be taken (Vinagre et al., 2003):

1. s^r, $(0 < r < 1)$, viewed as an operator, has a branch cut along the negative real axis for arguments of s on $(-\pi, \pi)$ but is free of poles and zeros.
2. It is well known that for interpolation or evaluation purposes, rational functions are sometimes superior to polynomials, roughly speaking, because of their ability to model functions with zeros and poles. In other words, for evaluation purposes, rational approximations frequently converge much more rapidly than PSE and have a wider domain of convergence in the complex plane.

In this section, for directly discretizing s^r, $(0 < r < 1)$, we shall concentrate on the IIR form of discretization where as a generating function we will adopt an Al-Alaoui idea on mixed scheme of Euler and Tustin operators (Al-Alaoui, 1993,

1997), but we will use a different ratio between both operators. The mentioned new operator, raised to power $\pm r$, has the form (Petráš, 2003a):

$$(\omega(z^{-1}))^{\pm r} = \left(\frac{1+a}{T}\frac{1-z^{-1}}{1+az^{-1}}\right)^{\pm r}, \tag{2.62}$$

where a is the ratio term and r is the fractional order. The ratio term a is the amount of phase shift and this tuning knob is sufficient for most engineering problems being solved.

In expanding the above in rational functions, we will use the CFE. It should be pointed out that for control applications, the obtained approximate discrete-time rational transfer function should be stable. Furthermore, for a better fit to the continuous frequency response, it would be of great interest to obtain discrete approximations with poles and zeros interlaced along the line $z \in (-1, 1)$ of the z plane. The direct discretization approximations proposed in this paper enjoy the desired properties.

The result of such approximation for an irrational function, $\widehat{G}(z^{-1})$, can be expressed by $G(z^{-1})$ in the CFE form (Vinagre et al., 2003):

$$\begin{aligned} G(z^{-1}) &\simeq a_0(z^{-1}) + \cfrac{b_1(z^{-1})}{a_1(z^{-1}) + \cfrac{b_2(z^{-1})}{a_2(z^{-1}) + \cfrac{b_3(z^{-1})}{a_3(z^{-1}) + \ldots}}} \\ &= a_0(z^{-1}) + \frac{b_1(z^{-1})}{a_1(z^{-1})+}\,\frac{b_2(z^{-1})}{a_2(z^{-1})+}\cdots\frac{b_3(z^{-1})}{a_3(z^{-1})+}\cdots, \end{aligned} \tag{2.63}$$

where a_i and b_i are either rational functions of the variable z^{-1} or constants. The application of the method yields a rational function, $G(z^{-1})$, which is an approximation of the irrational function $\widehat{G}(z^{-1})$.

The resulting discrete transfer function, approximating fractional-order operators, can be expressed as:

$$\begin{aligned} (\omega(z^{-1}))^{\pm r} &\approx \left(\frac{1+a}{T}\right)^{\pm r} \mathrm{CFE}\left\{\left(\frac{1-z^{-1}}{1+az^{-1}}\right)^{\pm r}\right\}_{p,q} \\ &= \left(\frac{1+a}{T}\right)^{\pm r} \frac{P_p(z^{-1})}{Q_q(z^{-1})}, \\ &= \left(\frac{1+a}{T}\right)^{\pm r} \frac{p_0 + p_1 z^{-1} + \cdots + p_m z^{-p}}{q_0 + q_1 z^{-1} + \cdots + q_n z^{-q}}, \end{aligned} \tag{2.64}$$

where $\mathrm{CFE}\{u\}$ denotes the continued fraction expansion of u; p and q are the orders of the approximation and P and Q are polynomials of degrees p and q. Normally, we can set $p = q = n$.

Modified and improved digital fractional-order differentiator using fractional sample delay and digital integrator using recursive Romberg integration rule and fractional order delay as well have been described in (Tseng and Lee, 2008).

Some others solutions for design of IIR approximation using least-squares, e.g. the Padé approximation, Prony's method and Shranks' method, were described in (Barbosa et al., 2006). The Prony and Shranks methods can give better approximations than widely used CFE method. The Padé and the CFE methods yield the same approximation (causal, stable and minimum phase). A different approach of the CFE method was used in (Maione, 2008).

Example 2.1. : Here we present some results for fractional order $r = \pm 0.5$ (half-order derivative/integral). The value of approximation order n is truncated to $n = 5$ and weighting factor a was chosen, $a = 1/3$. Assume sampling period $T = 0.001\,s$.

For $r = 0.5$ we have the following approximation of the fractional half-order derivative:

$$G(z^{-1}) = \frac{985.9 - 1315z^{-1} + 328.6z^{-2} + 36.51z^{-3}}{27 - 18z^{-1} - 3z^{-2} + z^{-3}}. \tag{2.65}$$

The Bode plots and unit step response of the digital fractional order differentiator (2.65) and the analytical continuous solution of a fractional semi-derivative are depicted in Fig. 2.3. Poles and zeros of the transfer function (2.65) lie in a unit circle.

For $r = -0.5$ we have the following approximation of the fractional half-order integral:

$$G(z^{-1}) = \frac{0.739 - 0.493z^{-1} - 0.0822z^{-2} + 0.0274z^{-3}}{27 - 36z^{-1} + 9z^{-2} + z^{-3}}. \tag{2.66}$$

The Bode plots and unit step response of the digital fractional-order integrator (2.66) and the analytical continuous solution of a fractional semi-derivative are depicted in Fig. 2.4. Poles and zeros of the transfer function (2.66) lie in a unit circle.

For simulation purpose, here we also present the Oustaloup's Recursive Approximation (ORA) algorithm (Oustaloup, 1995; Oustaloup et al., 2000). The method is based on the approximation of a function of the form:

$$H(s) = s^r, \quad r \in \mathrm{R}, \quad r \in [-1;1] \tag{2.67}$$

for the frequency range selected as (ω_b, ω_h) by a rational function:

$$\widehat{H}(s) = C_o \prod_{k=-N}^{N} \frac{s + \omega_k'}{s + \omega_k} \tag{2.68}$$

using the following set of synthesis formulas for zeros, poles and the gain:

$$\omega_k' = \omega_b \left(\frac{\omega_h}{\omega_b}\right)^{\frac{k+N+0.5(1-r)}{2N+1}} \qquad \omega_k = \omega_b \left(\frac{\omega_h}{\omega_b}\right)^{\frac{k+N+0.5(1-r)}{2N+1}} \qquad C_o = \left(\frac{\omega_h}{\omega_b}\right)^{-\frac{r}{2}} \prod_{k=-N}^{N} \frac{\omega_k}{\omega_k'},$$

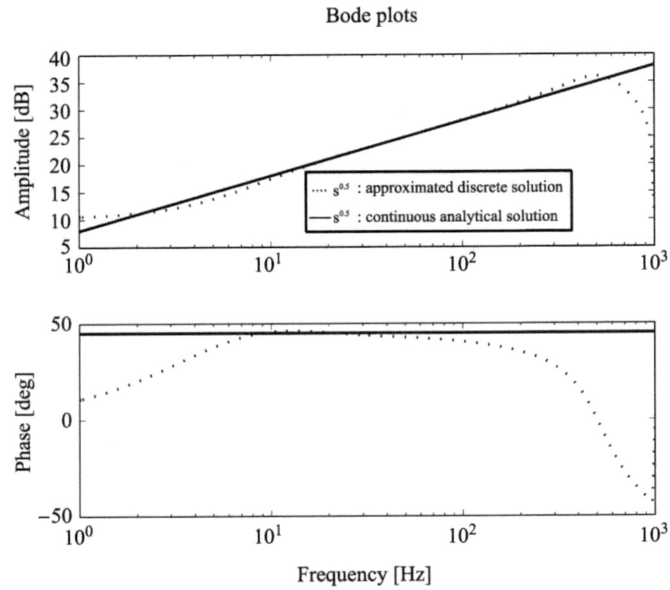

(a) Bode plots for $r = 0.5, n = 5, a = 1/3$, and $T = 0.001s$ in (2.64)

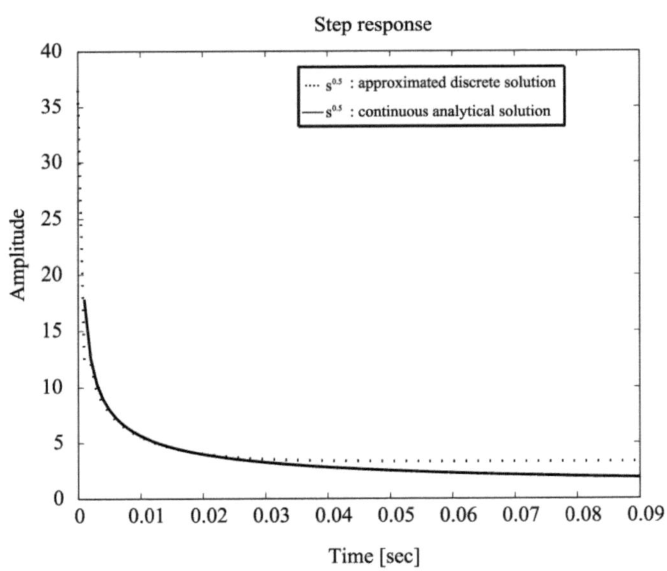

(b) Unit step responses for $r = 0.5, n = 5, a = 1/3$, and $T = 0.001s$ in (2.64)

Fig. 2.3 Characteristics of approximated fractional-order differentiator (2.65).

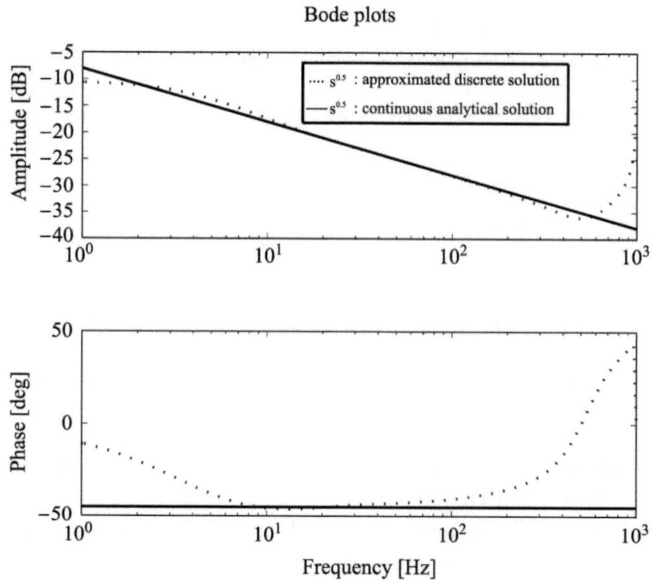

(a) Bode plots for $r = -0.5, n = 5, a = 1/3$, and $T = 0.001s$ in (2.64)

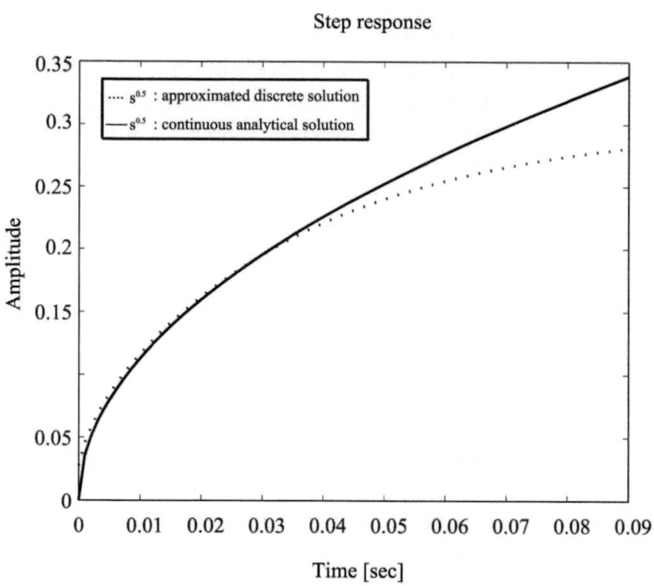

(b) Unit step responses for $r = -0.5, n = 5, a = 1/3$, and $T = 0.001s$ in (2.64)

Fig. 2.4 Characteristics of approximated fractional-order integrator (2.66).

where ω_h, ω_b are the high and low transitional frequencies. An implementation of this algorithm in Matlab as a function `ora_foc()` is given in (Chen, 2003).

Example 2.2. : Using the described Oustaloup's-Recursive-Approximation (ORA) method with

$$\omega_h = 10^3, \quad \omega_b = 10^{-3}, \tag{2.69}$$

the obtained approximation for fractional function $H(s) = s^{-\frac{1}{2}}$ is:

$$\widehat{H_5}(s) = \frac{s^5 + 74.97s^4 + 768.5s^3 + 1218s^2 + 298.5s + 10}{10s^5 + 298.5s^4 + 1218s^3 + 768.5s^2 + 74.97s + 1}. \tag{2.70}$$

The Bode plots and the unit step response of the approximated fractional-order integrator (2.70) are depicted in Fig. 2.5.

A description and overview of the various approximation methods and techniques (Carlson's (Carlson and Halijak, 1964), Charef's (Charef et al., 1992), CFE (Chen et al., 2006), CRONE-Oustaloup's (Oustaloup, 1995), descriptor approach (Tavazoei and Haeri, 2010), etc.) for continuous and discrete fractional-order models in form of IIR and FIR filters can be found in work (Vinagre et al., 2000). Besides mentioned methods, some other approaches were described in work (Podlubny et al., 2002). Last but not least, we should mention the approach proposed by Hwang, which is based on B-splines function (Hwang et al., 2002) and Podlubny's matrix approach (Podlubny, 2000; Podlubny et al., 2009).

The frequency domain approximation methods are not always reliable, especially in detecting chaos behavior in nonlinear systems (Tavazoei and Haeri, 2007a, 2008). As has been shown, due to error of approximation, numerical simulation may result in wrong conclusions, e.g. fake chaos is produced due to the implementation of the frequency domain approximation methods (Tavazoei and Haeri, 2007b). Simulation of the fractional-order system using the time domain methods is complicated and due to long memory characteristics of these systems requires a very long simulation time but on the other hand, it is more accurate. Applying some ideas as, for instance, short memory principle (Podlubny, 1999), we can reduce the computational cost of time-domain methods. Results obtained by these methods are more reliable than those determined using the frequency-based approximation (Tavazoei and Haeri, 2008).

Some of the mentioned frequency methods in both forms of approximation have been realized as the Matlab routines in Duarte Valerio's toolbox called *ninteger* (see detailed review in (Valerio, 2005)). In this toolbox was also created a Simulink block *nid* for fractional derivative and integral, where order of derivative/integral and method of its approximation can be selected. We will use this block for creating a model of fractional-order nonlinear systems in the Matlab/Simulink (see e.g. (Petráš, 2006)).

In Fig. 2.6 is depicted a screenshot of the *ninteger* Simulink block, where Oustaloup's approximation technique (CRONE) of order n was selected for the fractional-order 0.1 within desired frequency bandwidth.

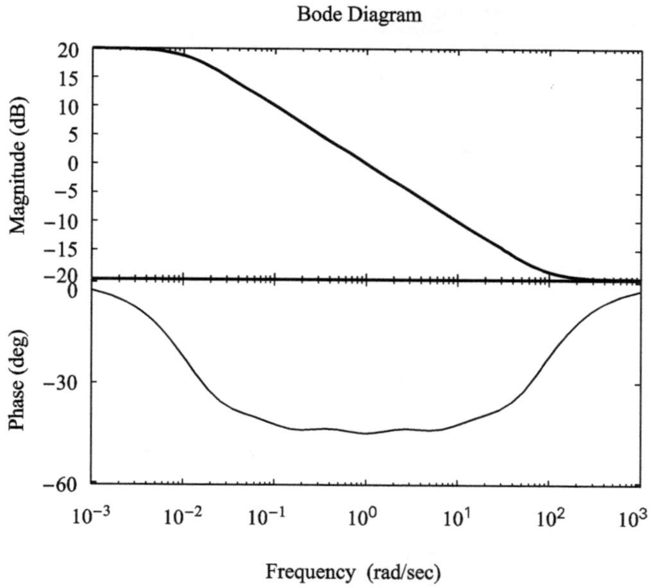

(a) Bode plots for $r = -0.5$ and $N = 5$

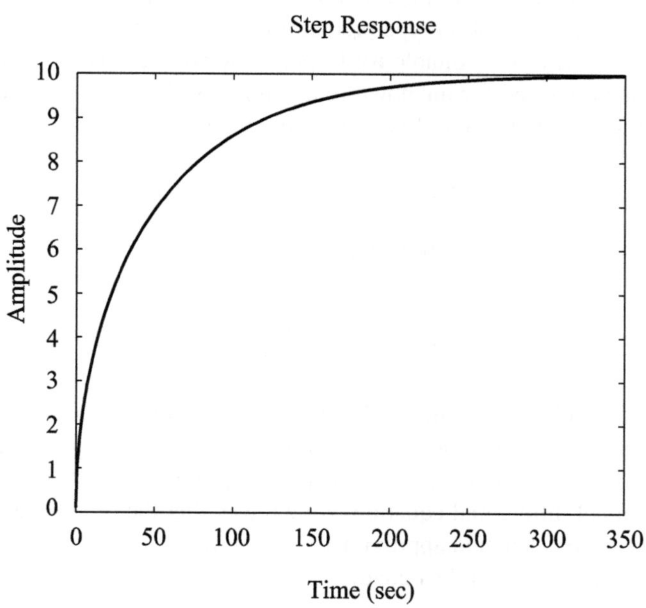

(b) Unit step response for $r = -0.5$ and $N = 5$

Fig. 2.5 Characteristics of approximated fractional-order integrator (2.70).

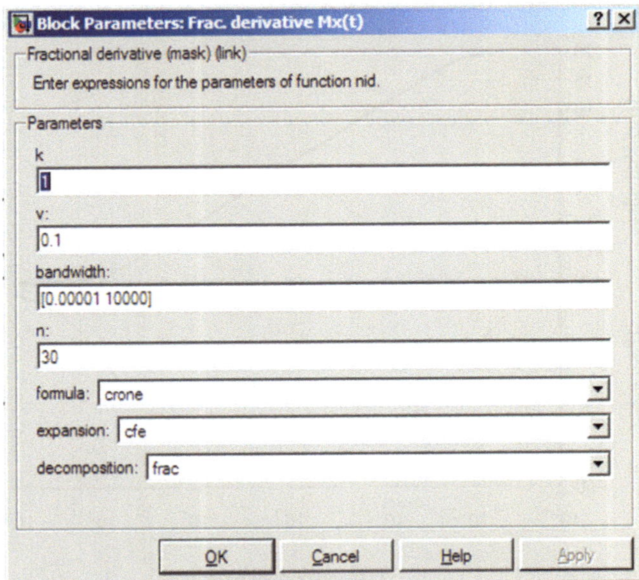

Fig. 2.6 Properties of Simulink *nid* block.

For numerical simulation of the fractional-order system a method on the basis of the Adams-Bashforth-Moulton type predictor-corrector scheme has also been proposed (Deng, 2007a). It is suitable for Caputo's derivative because it just requires the initial conditions and for unknown function it has clear physical meaning. The method is based on the fact that fractional differential equation

$$D_t^q y(t) = f(y(t),t), \; y^{(k)}(0) = y_0^{(k)}, k = 0,1,\ldots,m-1$$

is equivalent to the Volterra integral equation

$$y(t) = \sum_{k=0}^{[q]-1} y_0^{(k)} \frac{t^k}{k!} + \frac{1}{\Gamma(q)} \int_0^t (t-\tau)^{q-1} f(\tau, y(\tau)) d\tau. \tag{2.71}$$

Discretizing the Volterra equation (2.71) for uniform grid $t_n = nh$ $(n = 0,1,\ldots,N)$, $h = T_{sim}/N$ and using the short memory principle (fixed or logarithmic (Ford and Simpson, 2001)) we obtain a close numerical approximation of the true solution $y(t_n)$ of fractional differential equation while preserving the order of accuracy. Assume that we have calculated approximations $y_h(t_j)$, $j = 1,2,\ldots,n$ and we want to obtain $y_h(t_{n+1})$ by means of the equation

$$\begin{aligned} y_h(t_{n+1}) = & \sum_{k=0}^{m-1} \frac{t_{n+1}^k}{k!} y_0^{(k)} + \frac{h^q}{\Gamma(\alpha+2)} f(t_{n+1}, y_h^p(t_{n+1})) \\ & + \frac{h^q}{\Gamma(\alpha+2)} \sum_{j=0}^{n} a_{j,n+1} f(t_j, y_n(t_j)), \end{aligned} \tag{2.72}$$

where

$$a_{j,n+1} = \begin{cases} n^{q+1} - (n-q)(n+1)^q, & \text{if } j = 0, \\ (n-j+2)^{q+1} + (n-j)^{q+1} + 2(n-j+1)^{q+1}, & \text{if } 1 \le j \le n, \\ 1, & \text{if } j = n+1. \end{cases}$$

The preliminary approximation $y_h^p(t_{n+1})$ is called predictor and it is given by

$$y_h^p(t_{n+1}) = \sum_{k=0}^{m-1} \frac{t_{n+1}^k}{k!} y_0^{(k)} + \frac{1}{\Gamma(q)} \sum_{j=0}^{n} b_{j,n+1} f(t_j, y_n(t_j)), \tag{2.73}$$

where

$$b_{j,n+1} = \frac{h^q}{q}((n+1-j)^q - (n-j)^q). \tag{2.74}$$

A slightly improved predictor-corrector approach for solving the Fokker-Planck equation has been mentioned in (Deng, 2007b).

As shown in paper (Petráš, 2009), both mentioned time-domain numerical methods (Grünwald-Letnikov or Adams-Bashforth-Moulton) have approximately the same order of accuracy and good match of numerical solutions. A collection of various other numerical algorithms was also presented in (Diethelm et al., 2005).

2.10 Fractional Calculus and Electricity

There are a large number of electric and magnetic phenomena where the fractional calculus can be used (Westerlund, 2002). We will consider three of them — the capacitor, the inductor and the memristor.

Westerlund et al. in 1994, proposed a new linear capacitor model (Westerlund and Ekstam, 1994). It is based on Curie's empirical law of 1889 which states that the current through a capacitor is

$$I(t) = \frac{V_0}{h_1 t^\alpha},$$

where h_1 and α are constants, V_0 is the *dc* voltage applied at $t = 0$, and $0 < \alpha < 1$, $(\alpha \in \mathrm{R})$.

For a general input voltage $V(t)$ the current is

$$I(t) = C\frac{d^\alpha V(t)}{dt^\alpha} \equiv C_0 D_t^\alpha V(t), \tag{2.75}$$

where C is the capacitance of the capacitor. It is related to a kind of dielectric. Another constant α (order) is related to the losses of the capacitor. Westerlund provided in his work the table of various capacitor dielectrics with appropriate constant α which has been obtained experimentally by measurements.

For a current in the capacitor the voltage is

$$V(t) = \frac{1}{C}\int_0^t I(t)dt^{\alpha} \equiv \frac{1}{C}{}_0D_t^{-\alpha}I(t). \tag{2.76}$$

Then the impedance of a fractional capacitor is

$$Z_c(s) = \frac{1}{Cs^{\alpha}} = \frac{1}{\omega^{\alpha}C}e^{j(-\alpha\frac{\pi}{2})}. \tag{2.77}$$

Ideal Bode's characteristics of the transfer function for real capacitor (2.77) are depicted in Fig. 2.7.

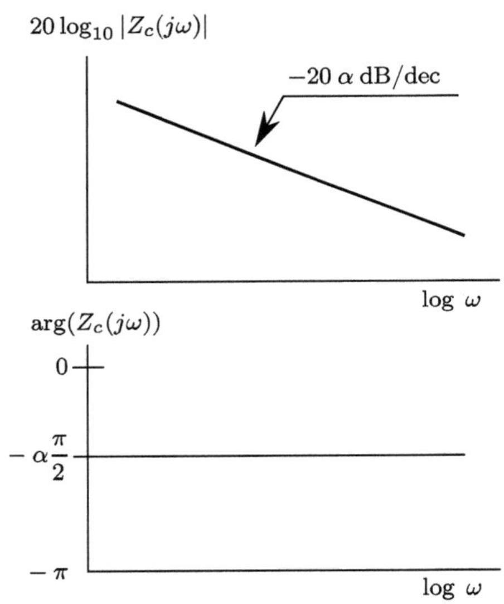

Fig. 2.7 Bode plots of real capacitor.

Westerlund in his work also described behavior of real inductor (Westerlund, 2002). For a general current in the inductor the voltage is

$$V(t) = L\frac{d^{\alpha}I(t)}{dt^{\alpha}} \equiv L_0D_t^{\alpha}I(t), \tag{2.78}$$

where L is the inductance of the inductor and constant α is related to the "proximity effect". Some coefficients for real inductors can be found in (Schafer and Kruger, 2008).

Then the impedance of a fractional inductor is

$$Z_L(s) = Ls^{\alpha} = \omega^{\alpha}Le^{j\alpha\frac{\pi}{2}}. \tag{2.79}$$

Ideal Bode's characteristics of the transfer function for real inductor (2.79) are depicted in Fig. 2.8.

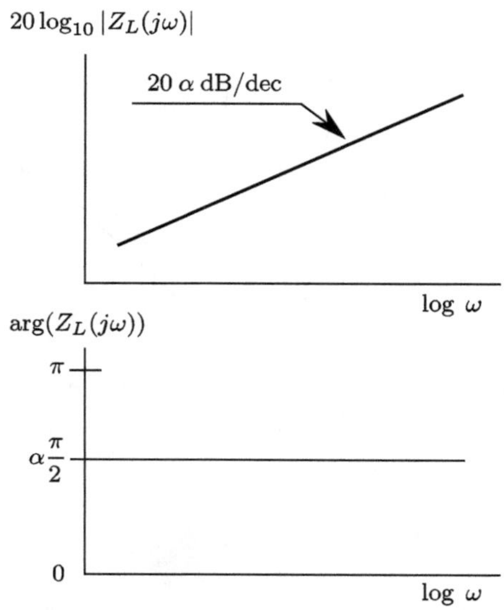

Fig. 2.8 Bode plots of real inductor.

Professor Leon O. Chua in 1971 predicted a new circuit element called memristor characterized by a relationship between the charge $q(t)$ and the flux $\phi(t)$. It is the fourth basic circuit element (Chua, 1971). The voltage across a charge – controlled memristor – is given by

$$v(t) = M(q(t))i(t), \quad \text{where} \quad M(q(t)) = d\phi(q)/dq. \tag{2.80}$$

The memristor used in this book is a flux – controlled memristor characterized by

$$I(t) = W(\phi(t))V(t), \quad \text{where} \quad W(\phi(t)) = dq(\phi)/d\phi, \tag{2.81}$$

where $W(\phi(t))$ is an incemental memductance of the memristor.

Professor Leon O. Chua and Dr. Sung–Mo Kang published another paper, in 1976, that described a large class of devices and systems they called memristive devices and systems (Chua and Kang, 1976). Whereas a memristor has mathematically scalar state, a system has vector state. The number of state variables is independent of, (usually greater than), the number of terminals. In that paper, Chua applied the model to empirically observed phenomena, including the Hodgkin-Huxley model of the axon and a thermistor at constant ambient temperature. He also described memristive systems in terms of energy storage and easily observed electrical characteristics. These characteristics match resistive random-access memory and phase-change memory, relating the theory to active areas of research. Chua extrapolated the conceptual symmetry between the resistor, inductor, and capacitor, and inferred that the memristor is a similarly fundamental device. Other scientists had already used fixed nonlinear flux-charge relationships, but Chua's theory introduces generality. This relation is illustrated in Fig. 2.9.

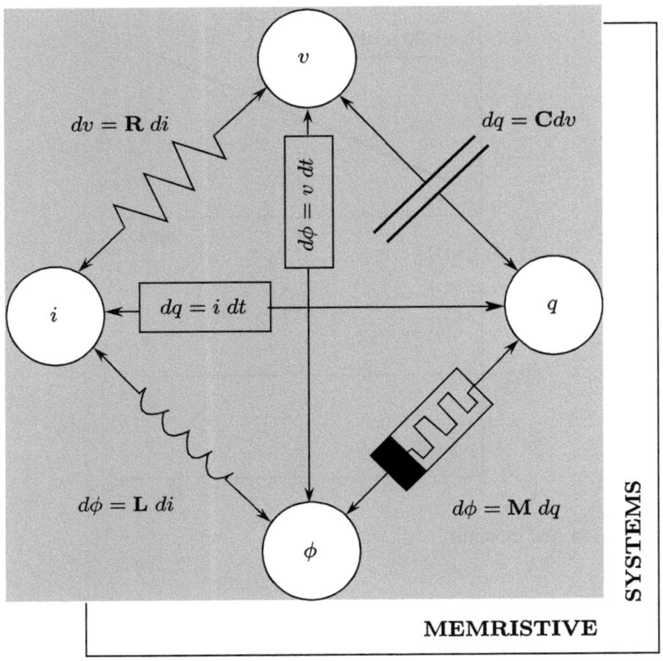

Fig. 2.9 Connection of four basic electrical elements (Inspired by figure in (Strukov et al., 2008)).

Possible applications of memristive systems (see e.g. (Wikipedia, 2009), etc.):

- new memory without access cycle limitations with new memory cells for more energy-efficient computers (Tour and He, 2008) e.g.: 1 bit = 1 memristor;
- new analog computers that can process and associate information in a manner similar to that of the human brain (Snider, 2008);
- new electronic circuits, e.g. (Itoh and Chua, 2009; Susse et al., 2005; Chua and Kang, 1976): cellular automata, voltage divider, switcher, compensator, AD – DA converters, etc.;
- new control systems/controllers with memory (Coopmans et al., 2009);

Noting from Faraday's law of induction that magnetic flux $\phi(t)$ is simply the time integral of voltage ($d\phi = V(t)\,dt$) and charge $q(t)$ is the time integral of current ($dq = I(t)\,dt$), the more convenient form of the current, voltage equation for the memristor, is (Chua, 1971).

$$M(q(t)) \int_0^t I(t)dt = \int_0^t V(t)dt, \tag{2.82}$$

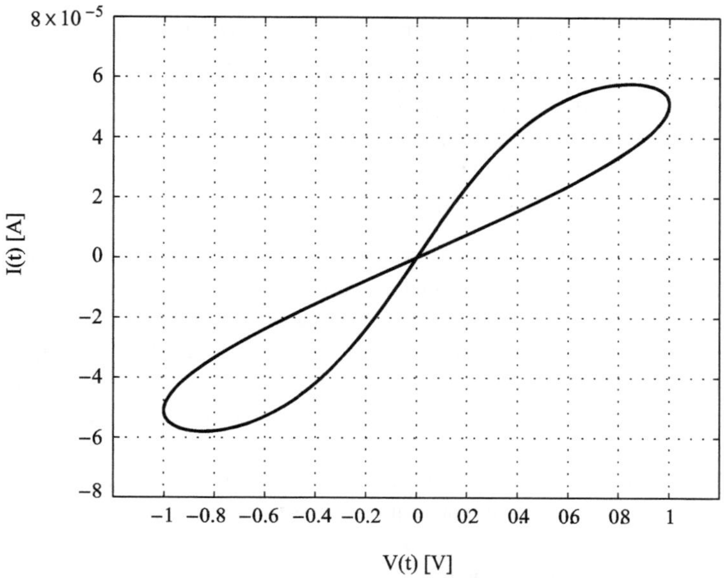

Fig. 2.10 Theoretical current-voltage characteristics of a memristor with applied voltage $V(t) = V_0(t)\sin(\omega t)$ for $V_0 = 1\,V$ and $\omega = 1\,rad/s$ (Thang, 2003).

where $M(q(t))$ is the memristance of the memristor. If $M(q(t))$ is a constant ($M(q(t)) \equiv R(t)$), then we obtain Ohm's law $R(t) = V(t)/I(t)$. If $M(q(t))$ is non-trivial, the equation is not equivalent because $q(t)$ and $M(q(t))$ will vary with time. Furthermore, the memristor is static if no current is applied. If $I(t) = 0$, we find $V(t) = 0$ and $M(t)$ is constant. This is the essence of the *memory effect*, which allows us to extend the notion of memristive systems to capacitive and inductive elements in the form of memcapacitors and meminductors whose properties depend on the state and history of the system (Mouttet, 2009; Ventra et al., 2009).

Similar to the capacitor and the inductor, the memristor is also not an ideal circuit element and we can predict the fractional-order model of such element. Applying the fractional calculus to relation (2.82), we obtain the following general formula for fractional-order memristive systems:

$$K\,{}_0D_t^{\gamma} I(t) = {}_0D_t^{\beta} V(t), \quad \gamma, \beta \in \mathrm{R}, \tag{2.83}$$

where K is the resistance, inductance, capacitance or memristance, respectively.

Aplying the Laplace transform technique (2.37), we get the following equation

$$K s^{\gamma} I(s) = s^{\beta} V(s) \tag{2.84}$$

and the resulting impedance of the memristive system (MS) is

$$Z_{MS}(s) = K s^{\gamma-\beta} = K s^{k}, \quad k \in \mathrm{R}, \tag{2.85}$$

where k is the real order of the memristive system and for ideal electrical elements has the following particular values, if

- $\gamma = 0$ and $\beta = 0$ then $k = 0$ – resistor and then $K = R\,[\Omega]$;
- $\gamma = -1$ and $\beta = 0$ then $k = -1$ – capacitor and then $K = 1/C\,[F]$;
- $\gamma = 0$ and $\beta = -1$ then $k = 1$ – inductor and then $K = L[H]$;
- $\gamma = -1$ and $\beta = -1$ then $k = 0$ – memristor and then $K = M(t)\,[\Omega]$.

However, as already has been mentioned, the real electrical elements are not ideal and with the help of fractional calculus it was shown that the intermediate cases between the known characteristic behavior of the electrical elements resistor R, capacitor C and inductor L change continuously (Susse et al., 2005). By deduction the memristor M, which has storage properties, could be also considered as a real electrical element with the fractional order of its mathematical model. The fractional calculus can help us to describe the memory behavior of the memristor. As we can see in Definitions (2.15), (2.18), and (2.21), the essence of the definitions consists of the memory term and takes into account the history. It is suitable for the description of memristive systems and their applications.

General characteristics of the transfer function of a real memristive system (2.85) are:

- Magnitude: constant slope of $\alpha 20dB/dec.$;
- Crossover frequency: a function of K;
- Phase: horizontal line of $\alpha\frac{\pi}{2}$;
- Nyquist: straight line at argument $\alpha\frac{\pi}{2}$.

The above concepts of memory devices are not necessarily limited to resistance – memristor, but can in fact be generalized to capacitative and inductive systems. If $x(t)$ denotes a set of n state variables describing the internal state of the system, $u(t)$ and $y(t)$ are any two complementary constitutive variables (current, charge, voltage, or flux) denoting input and output of the system, and g is a generalized response, we can define a general class of nth-order u-controlled dynamical systems called memristive systems or devices described by the following equations (Chua and Kang, 1976)

$$\begin{aligned} \frac{dx(t)}{dt} &= f(x,u,t), \\ y(t) &= g(x,u,t)u(t), \end{aligned} \tag{2.86}$$

where f is a continuous n-dimensional vector function and we assume unique solution for any initial state $x(t)$ at time $t = t_0$.

General fractional-order differential equation (2.83) can be rewritten to its canonical form and then Eqs. (2.86) become

$$\begin{aligned} {}_0D_t^{\alpha} x(t) &= f(x,u,t) \\ y(t) &= g(x,u,t)u(t). \end{aligned} \tag{2.87}$$

An alternative version of the connection diagram in Fig. 2.9 is presented in Fig. 2.11. By moving parts of the original diagram (moving the resistor symbol

R to the center, for example), and assuming continuous real numerical space for α and β, space is included for the "*Fractor*" as presented by Bohannan (Bohannan, 2002), as well as a new "element", a so-called fractional inductor, or "*Fractductor*". This device has a fractional-order coupling between flux and current.

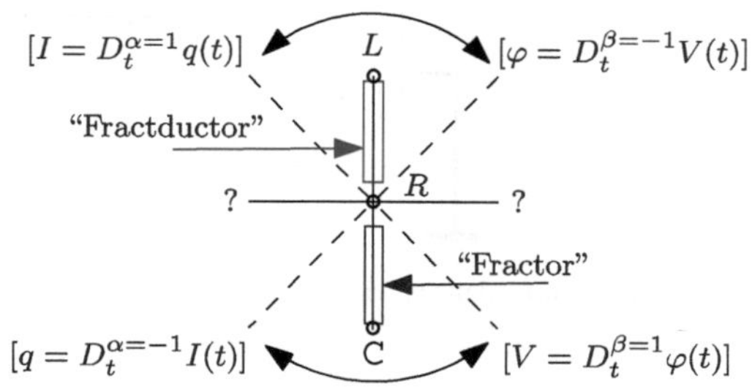

Fig. 2.11 Alternative electrical component connection diagram.

2.10.1 Analogue Fractional-Order Circuits

We are able to define arbitrary real order α for the memristive system behavior description (2.85). The amplitude of this impedance function is $A = 20\alpha$ and the phase angle is $\Phi = \alpha(\pi/2)$ for $\alpha \in \mathrm{R}$. Electrical elements (memristive system or fractance) with such property are sometimes called constant phase elements for certain frequency range (Bode, 1949). So far, the constant phase elements (CPE) have been approximated by the ladder network constructed from RLC elements, tree network, metal-insulator-liquid interface, etc. (see e.g. (Biswas et al., 2006; Carlson and Halijak, 1964; Charef, 2006; Dutta Roy, 1967; Ichise et al., 1971; Jones and Shenoi, 1970; Krishna and Reddy, 2008; Nakagava and Sorimachi, 1992; Oldham and Zoski, 1983; Petráš et al., 2002; Takyar and Georgiu, 2007; Wang, 1987), etc.).

We can use an active operating amplifier (op-amp) and its inverting connection with impedance Z_1 in direct connection and impedance Z_2 in feedback connection. The transfer function of the circuit depicted in Fig. 2.12 is

$$H(s) = \frac{V_o(s)}{V_{in}(s)} = -\frac{Z_2(s)}{Z_1(s)}.$$

Generally, basic electrical elements (resistor, capacitor, inductor, memristor), electrical networks (RC ladder, RC tree, RLC grid, CPE) or new electrical elements (fractor, fractductor) can be used as electrical elements with the impedances Z_1 and Z_2. In this way we can obtain various new dividers, filters, integrators, etc.

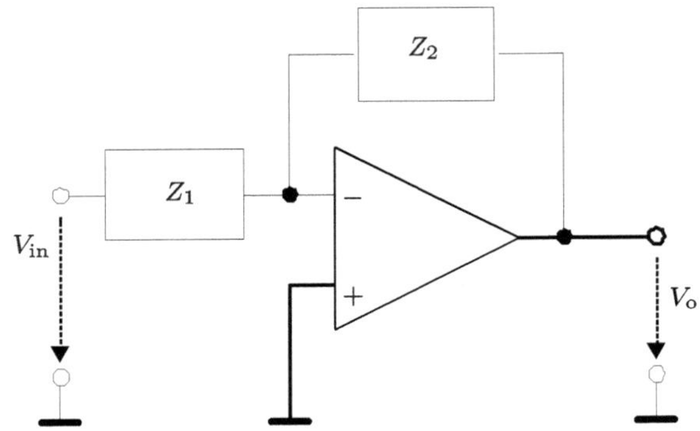

Fig. 2.12 Basic connection of two impedances with op-amp.

2.10.2 Experimental Measurement

Here we present a measurement on a new "element", a so-called fractional inductor, or "*Fractductor*". This device has a fractional-order coupling between flux and current.

Preliminary attempts to construct a fractductor have produced a device using magnetorheological fluid as the core in a transformer-like device. A bode plot (Fig. 2.13), and basic block schematic (Fig. 2.14) are shown (Petráš et al., 2009).

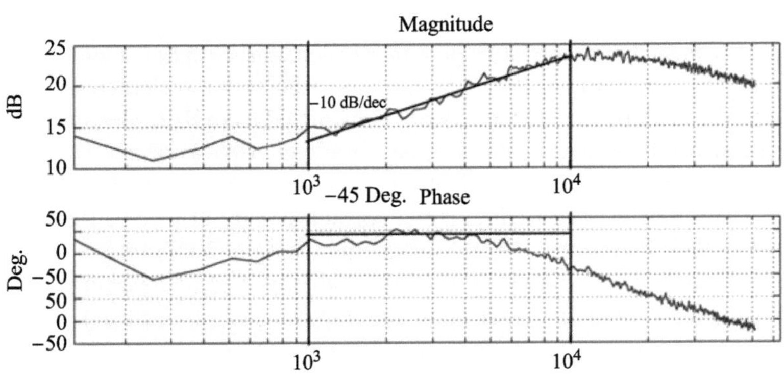

Fig. 2.13 Bode plots of an experimental fractional flux coupling device, the fractductor.

As we can see in Fig. 2.14, connections of electrical elements were done according to suggestions described in prevoius subsections. It is a practical realization of fractional-order memristive system which can be used for fractional-order controller implementations as well. Estimated fractional order of the real fractductor is $\alpha = 0.5$ (see Bode plots in Fig. 2.13).

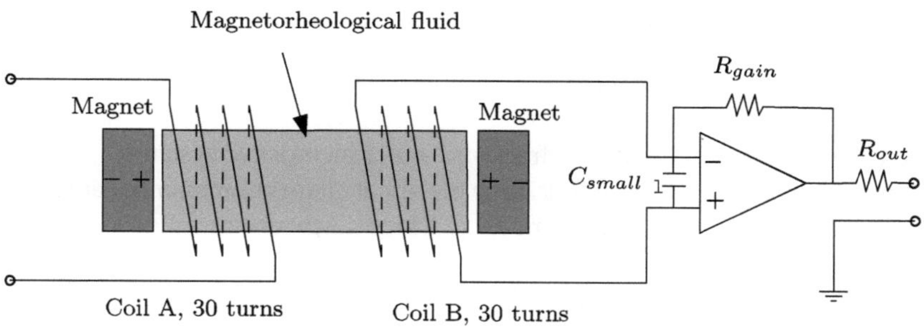

Fig. 2.14 Simplified schematic diagram of the fractductor and test circuit.

2.10.3 Additional Remarks

One can say that the memristor is not a basic electrical element, being nonlinear because of its nonlinear current-voltage characteristic (Fig. 2.10), and the other basic elements (resistor, capacitor and inductor) are linear elements.

Basically, all resistors are in reality nonlinear. Current which flows through the resistors causes their heating and therefore their conductivity grows. The characteristic is $V = f(R,I)$. The situation is similar also to capacitors. Permittivity of some dielectrics varies with the magnetic field intensity and therefore capacity is a nonlinear function of voltage on the capacitor. The current that flows across the capacitor is given by $I = d(CV)/dt$, where capacitance C is variable and is a function of voltage V. For inductors, their nonlinearity is given by hysteresis. Permeability of inductor cores depends on the magnetic field and therefore inductivity is a function of current. The voltage on nonlinear inductor is $U = d(LI)/dt$, where L is inductivity varying with current I. According to above notes, we can state that all basic elements are nonlinear but for some of them the nonlinearity is very low and negligible (Kotek et al., 1973). For this reason we can consider them to be linear and use much easier mathematical tools for their modeling. In fact, this consideration indirectly proves that the memristor is basic electrical element, the so-called *fourth element.*

In general, if we consider electromechanical analogy, we can use a fractional-order model for viscoelastic materials, where there is a relationship between stress and strain for solids. There are several kinds of such models which have been described in literature (Podlubny, 1999). For our purpose a generalized five-parameter Zener model is suitable:

$$\sigma(t) + a_1 D_t^{\alpha}\sigma(t) = b_0\varepsilon(t) + b_1 D_t^{\beta}\varepsilon(t),$$

where $\sigma(t)$ is the stress and $\varepsilon(t)$ is the strain.

The proposed model of memristive systems in the form

$$K\, D_t^{\gamma} I(t) = D_t^{\beta} V(t), \quad \gamma, \beta \in \mathrm{R}$$

could be modified according to the Zener model to

$$b_0 I(t) + b_1 D_t^{\gamma} I(t) = a_0 V(t) + a_1 D_t^{\beta} V(t), \quad \gamma, \beta \in \mathrm{R}, \tag{2.88}$$

which is a more general model of fractional-order memristive systems.

Taking into account that all four basic electrical elements are nonlinear and fractional, we can modify the general model (2.88) to a new one as follows:

$$f_0 I(t) + f_1 D_t^{\gamma} I(t) = f_2 V(t) + f_3 D_t^{\beta} V(t), \quad \gamma, \beta \in \mathrm{R}, \tag{2.89}$$

where f_0, f_1, f_2, and f_3 are nonlinear functions of R, C, L, and M depending on voltage $V(t)$, current $I(t)$, charge $q(t)$ and flux $\phi(t)$ at time t. It can be written as $f_{0,1,2,3} = \mathbf{F}(R(t), C(t), L(t), M(t), V(t), I(t), q(t), \phi(t), t)$. The new mathematical model (2.89) should provide a satisfactory description of most real electrical elements.

References

Al-Alaoui M. A., 1993, Novel digital integrator and differentiator, *Electron. Lett.*, **29**, 376–378.

Al-Alaoui M. A., 1997, Filling the gap between the bilinear and the backward difference transforms: An interactive design approach, *Int. J. Elect. Eng. Edu.*, **34**, 331–337.

Barbosa R. S., Machado J. A. T. and Silva M. F., 2006, Time domain design of fractional differintegrators using least-squares, *Signal Processing*, **86**, 2567–2581.

Bode H. W., 1949, *Network Analysis and Feedback Amplifier Design*, Tung Hwa Book Company, Shanghai.

Bohannan G., 2002, Analog realization of a fractional control element – Revised, *Proc. of the 41st IEEE Int. Conf. on Decision and Control*, December 9, Las Vegas, USA.

Biswas K., Sen S. and Dutta P. K., 2006, Realization of a constant phase element and its performance study in a differentiator circuit, *IEEE Transactions on Circuits and Systems II: Express Briefs*, **53**, 802–806.

Caputo M., 1967, Linear models of dissipation whose Q is almost frequency independent, *Geophys. J. R. Astr. Soc.*, **13**, 529–539. (Reprinted recently in: *Fract. Calc. Appl. Anal.*, **11**, 3–14, 2008.)

Carlson G. E. and Halijak C. A., 1964, Approximation of fractional capacitors $(1/s)^{1/n}$ by a regular Newton process, *IEEE Trans. on Circuit Theory*, **11**, 210–213.

Coopmans C., Petráš I. and Chen Y. Q. , 2009, Analogue fractional–order generalized memristive devices, *Proc. of the ASME 2009 International Design Engineering Technical Conferences & Computers and Information in Engineering*

Conference, DETC2009/MSNDC-86861, August 30 – September 2, San Diego, USA.

Charef A., Sun H. H., Tsao Y. Y. and Onaral B., 1992, Fractal system as represented by singularity function, *IEEE Transactions on Automatic Control*, **37**, 1465–1470.

Charef A., 2006, Modeling and analog realization of the fundamental linear fractional order differential equation, *Nonlinear Dynamics*, **46**, 195–210.

Chen Y. Q., 2003, Oustaloup-Recursive-Approximation for Fractional Order Differentiators. MathWorks, Inc., FileExchage, http://www.mathworks.com/matlabcentral / fileexchange / 3802-outstaloup-recursive-appriximation-for-fractional-order-differentiators

Chen Y. Q., Ahn H. S. and Xue D., 2006, Robust controllability of interval fractional order linear time invariant systems, *Signal Processing*, **86**, 2794–2802.

Chua L. O., 1971, Memristor – the missing circuit element, *IEEE Transaction on Circuit Theory*, **CT-18**, 507–519.

Chua L. O. and Kang S. M., 1976, Memristive devices and systems, *Proceedings of the IEEE*, **64**, 209–223.

Deng W., 2007a, Short memory principle and a predictor-corrector approach for fractional differentional equations, *Journal of Computational and Applied Mathematics*, **206**, 174–188.

Deng W., 2007b, Numerical algorithm for the time fractional Fokker-Planck equation, *Journal of Computational Physics*, **227**, 1510–1522.

Diethelm K., Ford N. J., Freed A. D. and Luchko Yu., 2005, Algorithms for the fractional calculus: A selection of numerical methods, *Comput. Methods Appl. Mech. Engrg.*, **194**, 743–773.

Djrbashian M. M., 1993, *Harmonic Analysis and Boundary Value problems in the Complex Domain*, Birkhäuser, Basel.

Dorčák Ľ., 1994, Numerical Models for the Simulation of the Fractional-Order Control Systems, *UEF-04-94, The Academy of Sciences, Inst. of Experimental Physic*, Košice, Slovakia.

Dutta Roy S. C., 1967, On the realization of a constant-argument immitance of fractional operator, *IEEE Trans. on Circuit Theory*, **14**, 264–374.

Ford N. and Simpson A., 2001, The numerical solution of fractional differential equations: speed versus accuracy, *Num. Anal. Report 385*, Manchester Centre for Computational Mathematics.

Gorenflo R., Kilbas A. A. and Rogosin S. V., 1998, On the generalized mittag-leffler type functions, *Integral Transform. Spec. Funct.*, **7**, 215–224.

Heymans N. and Podlubny I., 2006, Physical interpretation of initial conditions for fractional differential equations with Riemann-Liouville fractional derivatives, *Rheologica Acta*, **45**, 765–772.

Hwang C., Leu J. F. and Tsay S. Y., 2002, A note on time-domain simulation of feedback fractional-order systems, *IEEE Transactions on Automatic Control*, **47**, 625–631.

Ichise M., Nagayanagi Y. and Kojima T., 1971, An analog simulation of non-integer order transfer functions for analysis of electrode processes, *J. Electroanal. Chem.*, **33**, 253–265.

Itoh M. and Chua L. O., 2009, Memristor cellular automata and memristor discrete-time cellular neural networks, *International Journal of Bifurcation and Chaos*, **19**, 3605–3656.

Jones H. E. and Shenoi B. A., 1970, Maximally flat lumped-element approximation to fractional operator immitance function, *IEEE Trans. on Circuit and Systems*, **17**, 125–128.

Kotek Z., Kubik S. and Razim M., 1973, *Nonlinear Dynamical Systems*, TKI-SNTL, Praha (in Czech).

Krishna B. T. and Reddy K. V. V. S., 2008, Active and Passive Realization of Fractance Device of Order 1/2, *Active and Passive Electronic Components*, Hindawi Publishing Corporation, Article ID **369421**, DOI: 10.1155/2008/369421.

Lubich C., 1986, Discretized fractional calculus, *SIAM J. Math. Anal.*, **17**, 704–719.

Maione G., 2008, Continued fractions approximation of the impulse response of fractional-order dynamic systems, *IET Control Theory Appl.*, **2**, 564–572.

Miller K. S. and Ross B., 1993, *An Introduction to the Fractional Calculus and Fractional Differential Equations*, John Wiley & Sons. Inc., New York.

Mittag-Leffler M. G., 1903, Sur la nouvelle fonction $E_\alpha(x)$, *Comptes Rendus*, Acad. Sci. Paris, **137**, 554–558.

Mouttet B., 2009, An Introduction to Memimpedance and Memadmittance Systems Analysis, http://knol.google.com/k/blaise-mouttet/.

Nakagava M. and Sorimachi K., 1992, Basic characteristics of a fractance device, *IEICE Trans. Fundamentals*, **E75-A**, 1814–1818.

Oldham K. B. and Spanier J., 1974, *The Fractional Calculus*, Academic Press, New York.

Oldham K. B. and Zoski C. G., 1983, Analogue instrumentation for processing polarographic data, *J. Electroanal. Chem.*, **157**, 27–51.

Oustaloup A., 1995, *La Derivation Non Entiere: Theorie, Synthese et Applications*, Hermes, Paris.

Oustaloup A., Levron F., Mathieu B. and Nanot F. M., 2000, Frequency-band complex noninteger differentiator: characterization and synthesis, *IEEE Trans. on Circuits and Systems I: Fundamental Theory and Applications I*, **47**, 25–39.

Petráš I., Podlubny I., O'Leary P., Dorčák Ľ. and Vinagre B. M., 2002, *Analogue Realization of Fractional Order Controllers*, Faculty of BERG, Technical University of Kosice.

Petráš I., 2003a, Digital Fractional Order Differentiator/integrator - IIR type. MathWorks, Inc., http://www.mathworks.com/matlabcentral/fileexchange/3672.

Petráš I., 2003b, Digital Fractional Order Differentiator/integrator - FIR type. MathWorks, Inc., http://www.mathworks.com/matlabcentral/fileexchange/3673.

Petráš I., 2006, Method for simulation of the fractional order chaotic systems, *Acta Montanistica Slovaca*, **11**, 273–277.

Petráš I., 2009, Chaos in the fractional-order Volta's system: modeling and simulation, *Nonlinear Dynamics*, **57**, 157–170.

Petráš I., Chen Y. Q. and Coopmans C., 2009, Fractional-order memristive systems, *Proc. of the 14th International Conference on Emerging Technologies and Factory Automation - ETFA 2009*, September 22–26, Mallorca.

Podlubny I., 1999, *Fractional Differential Equations*, Academic Press, San Diego.

Podlubny I., 2000, Matrix approach to discrete fractional calculus, *Fractional Calculus and Applied Analysis*, **3**, 359–386.

Podlubny I., 2002, Geometric and physical interpretation of fractional integration and fractional differentiation, *Fractional Calculus and Applied Analysis*, **5**, 367–386.

Podlubny I., Petráš, I., Vinagre, B. M., O'Leary P., Dorčák Ľ., 2002, Analogue Realizations of Fractional-Order Controllers, *Nonlinear Dynamics*, **29**, 281–296.

Podlubny I. and Kacenak M., 2005, Mittag-Leffler function. MathWorks, Inc., http://www.mathworks.com/matlabcentral/fileexchange/8738.

Podlubny I., Chechkin A., Škovránek T., Chen Y. Q. and Vinagre B. M. J., 2009, Matrix approach to discrete fractional calculus II: Partial fractional differential equations, *Journal of Computational Physics*, **228**, 3137–3153.

Schafer I. and Kruger K., 2008, Modelling of lossy coils using fractional derivatives, *J. Phys. D: Appl. Phys.*, **41**, 1–8.

Snider G. S., 2008, Spike-timing-dependent learning in memristive nanodevices, *Proc. of the IEEE International Symposium on Nanoscale Architectures*, NANOARCH 2008, June 12–13, 85–92.

Strukov D. B., Snider G. S., Stewart D. R. and Williams R. S., 2008, The missing memristor found, *Nature*, **453**, 80–83.

Susse R., Domhardt A. and Reinhard M., 2005, Calculation of electrical circuits with fractional characteristics of construction elements, *Forsch Ingenieurwes*, **69**, 230–235.

Takyar M. S. and Georgiu T. T., 2007, The fractional integrator as a control design element, *Proceedings of the 46th IEEE Conference on Decision and Control*, New Orleans, USA, Dec. 12–14, 239–244.

Tavazoei M. S. and Haeri M., 2007a, Unreliability of frequency-domain approximation in recognising chaos in fractional-order systems, *IET Signal Proc.*, **1**, 171–181.

Tavazoei M. S. and Haeri, M., 2007b, A necessary condition for double scroll attractor existence in fractional – order systems, *Physics Letters A*, **367**, 102–113.

Tavazoei M. S. and Haeri M., 2008, Limitations of frequency domain approximation for detecting chaos in fractional order systems, *Nonlinear Analysis*, **69**, 1299–1320.

Tavazoei M. S. and Haeri M., 2010, Rational approximations in the simulation and implementation of fractional-order dynamics: A descriptor system approach, *Automatica*, **46**, 94–100.

Thang H. M., 2003, Memristor model, MathWorks, Inc., FileExchage, http://www.mathworks.com/matlabcentral/fileexchange/25082.

Tour J. M. and He T., 2008, The fourth element, *Nature*, **453**, 42–43.

Tseng C. C. and Lee S. L., 2008, Digital IIR integrator design using recursive Rombĕrg iňtegration rule and fractional sample delay, *Signal Processing*, **88**, 2222–2233.

Valerio D., 2005, Toolbox `ninteger` for Matlab, v. 2.3 (September, 2005), http://web.ist.utl.pt/duarte.valerio/ninteger/ninteger.htm.

Ventra M. D., Pershin Y. V. and Chua L. O., 2009, Circuit elements with memory: memristors, memcapacitors and meminductors, e-print arxive: http://arxiv.org/abs/0901.3682.

Vinagre B. M., Podlubny I., Hernández A. and Feliu V., 2000, Some approximations of fractional order operators used in control theory and applications, *Fractional Calculus and Applied Analysis*, **3**, 231–248.

Vinagre B. M., Chen Y. Q. and Petráš I., 2003, Two direct Tustin discretization methods for fractional-order differentiator/integrator, *J. Franklin Inst.*, **340**, 349–362.

Wang J. C., 1987, Realizations of generalized Warburg impedance with RC ladder networks and transmission lines, *Journal of The Electrochemical Society*, **134**, 1915–1920.

Westerlund S. and Ekstam L., 1994, Capacitor theory, *IEEE Trans. on Dielectrics and Electrical Insulation*, **1**, 826–839.

Westerlund S., 2002, *Dead Matter Has Memory!* Causal Consulting, Kalmar, Sweden.

Wikipedia, 2009, "Memristor", Wikimedia Foundation, Inc. (free encyclopedia), http://en.wikipedia.org/wiki/Memristor.

Chapter 3
Fractional-Order Systems

3.1 Fractional LTI Systems

A general fractional-order system can be described by a fractional differential equation of the form

$$\begin{aligned} & a_n D^{\alpha_n} y(t) + a_{n-1} D^{\alpha_{n-1}} y(t) + \cdots + a_0 D^{\alpha_0} y(t) \\ & = b_m D^{\beta_m} u(t) + b_{m-1} D^{\beta_{m-1}} u(t) + \cdots + b_0 D^{\beta_0} u(t), \end{aligned} \tag{3.1}$$

where $D^\gamma \equiv {}_0D_t^\gamma$ denotes the Grünwald-Letnikov, the Riemann-Liouville or the Caputo's fractional derivative (Podlubny, 1999a). The corresponding transfer function of *incommensurate* real orders has the following form (Podlubny, 1999a):

$$G(s) = \frac{b_m s^{\beta_m} + \cdots + b_1 s^{\beta_1} + b_0 s^{\beta_0}}{a_n s^{\alpha_n} + \cdots + a_1 s^{\alpha_1} + a_0 s^{\alpha_0}} = \frac{Q(s^{\beta_k})}{P(s^{\alpha_k})}, \tag{3.2}$$

or in the frequency domain it has form (Petráš et al., 2000):

$$G(j\omega) = \frac{b_m (j\omega)^{\beta_m} + \cdots + b_1 (j\omega)^{\beta_1} + b_0 (j\omega)^{\beta_0}}{a_n (j\omega)^{\alpha_n} + \cdots + a_1 (j\omega)^{\alpha_1} + a_0 (j\omega)^{\alpha_0}} = \frac{Q((j\omega)^{\beta_k})}{P((j\omega)^{\alpha_k})}, \tag{3.3}$$

where a_k $(k = 0, \ldots, n)$, b_k $(k = 0, \ldots, m)$ are constants, and α_k $(k = 0, \ldots, n)$, β_k $(k = 0, \ldots, m)$ are arbitrary real or rational numbers and without loss of generality they can be arranged as $\alpha_n > \alpha_{n-1} > \cdots > \alpha_0$, and $\beta_m > \beta_{m-1} > \cdots > \beta_0$.

The incommensurate order system (3.2) can also be expressed in commensurate form by the multivalued transfer function (Bayat and Afshar, 2008)

$$H(s) = \frac{b_m s^{m/v} + \cdots + b_1 s^{1/v} + b_0}{a_n s^{n/v} + \cdots + a_1 s^{1/v} + a_0}, \quad v > 1. \tag{3.4}$$

Note that every fractional-order system can be expressed in the form (3.4) and the domain of the $H(s)$ definition is a Riemann surface with v Riemann sheets (LePage, 1961).

In the particular case of *commensurate* order systems, it holds that $\alpha_k = \alpha k, \beta_k = \alpha k, 0 < \alpha < 1, \forall k \in Z$, and the transfer function has the following form:

$$G(s) = K_0 \frac{\sum_{k=0}^{M} b_k (s^\alpha)^k}{\sum_{k=0}^{N} a_k (s^\alpha)^k} = K_0 \frac{Q(s^\alpha)}{P(s^\alpha)}. \tag{3.5}$$

With $N > M$, the function $G(s)$ becomes a proper rational function in the complex variable s^α which can be expanded in partial fractions of the following form:

$$G(s) = K_0 \left[\sum_{i=1}^{N} \frac{A_i}{s^\alpha + \lambda_i} \right], \tag{3.6}$$

where $\lambda_i \, (i = 1, 2, \ldots, N)$ are the roots of the pseudo-polynomial $P(s^\alpha)$ or the system poles which are assumed to be simple without loss of generality. The analytical solution of the system (3.6) can be expressed as

$$y(t) = L^{-1} \left\{ K_0 \left[\sum_{i=1}^{N} \frac{A_i}{s^\alpha + \lambda_i} \right] \right\} = K_0 \sum_{i=1}^{N} A_i t^\alpha E_{\alpha,\alpha}(-\lambda_i t^\alpha), \tag{3.7}$$

where $E_{\mu,\nu}(z)$ is the Mittag-Leffler function defined as (2.3).

A fractional-order plant to be controlled can be described by a typical n-term linear homogeneous fractional-order differential equation (FODE) in time domain

$$a_n D_t^{\alpha_n} y(t) + \cdots + a_1 D_t^{\alpha_1} y(t) + a_0 D_t^{\alpha_0} y(t) = 0, \tag{3.8}$$

where $a_k \, (k = 0, 1, \ldots, n)$ are constant coefficients of the FODE; $\alpha_k \, (k = 0, 1, 2, \ldots, n)$ are real numbers. Without loss of generality, assume that $\alpha_n > \alpha_{n-1} > \cdots > \alpha_0 \geq 0$.

The analytical solution of the FODE (3.8) is given by general formula in the form (Podlubny, 1999a):

$$\begin{aligned} y(t) = & \frac{1}{a_n} \sum_{m=0}^{\infty} \frac{(-1)^m}{m!} \sum_{\substack{k_0+k_1+\cdots+k_{n-2}=m \\ k_0 \geq 0; \ldots; k_{n-2} \geq 0}} (m; k_0, k_1, \ldots, k_{n-2}) \\ & \times \prod_{i=0}^{n-2} \left(\frac{a_i}{a_n} \right)^{k_i} \mathscr{E}_m\Big(t, -\frac{a_{n-1}}{a_n}; \alpha_n - \alpha_{n-1}, \alpha_n + \sum_{j=0}^{n-2} (\alpha_{n-1} - \alpha_j) k_j + 1\Big), \end{aligned} \tag{3.9}$$

where $(m; k_0, k_1, \ldots, k_{n-2})$ are the multinomial coefficients and $\mathscr{E}_k(t, \lambda; \mu, \nu)$ is the function of Mittag-Leffler type introduced by Podlubny (Podlubny, 1999a). The function is defined by

$$\mathscr{E}_k(t, \lambda; \mu, \nu) = t^{\mu k + \nu - 1} E_{\mu,\nu}^{(k)}(\lambda t^\mu), \quad k = 0, 1, 2, \ldots \tag{3.10}$$

where $E^{(k)}_{\mu,\nu}(z)$ is k-th derivative of the Mittag-Leffler function of two parameters given by

$$E^{(k)}_{\mu,\nu}(z) = \sum_{i=0}^{\infty} \frac{(i+k)!\, z^i}{i!\, \Gamma(\mu i + \mu k + \nu)}, \qquad k = 0, 1, 2, \ldots. \tag{3.11}$$

The Laplace transform of the function $\mathscr{E}_k(t, \pm\lambda; \alpha, \beta)$ is (Podlubny, 1999a):

$$L\{\mathscr{E}_k(t, \pm\lambda; \alpha, \beta)\} = \frac{k!\, s^{\alpha-\beta}}{(s^\alpha \mp \lambda)^{k+1}}$$

for $s > |\lambda|^{1/\alpha}$.

The Laplace transforms for several other Mittag-Leffler type functions are summarized as follows (Gorenflo et al., 2004; Magin et al., 2009; Podlubny, 1999a):

$$\begin{aligned} L\{E_\alpha(-\lambda t^\alpha)\} &= \frac{s^{\alpha-1}}{s^\alpha + \lambda}, \\ L\{t^{\alpha-1} E_{\alpha,\alpha}(-\lambda t^\alpha)\} &= \frac{1}{s^\alpha + \lambda}, \\ L\{t^{\beta-1} E_{\alpha,\beta}(-\lambda t^\alpha)\} &= \frac{s^{\alpha-\beta}}{s^\alpha + \lambda} \end{aligned} \tag{3.12}$$

for $s > |\lambda|^{1/\alpha}$.

A useful list of Laplace and inverse Laplace transforms of functions related to fractional calculus is presented in Appendix B and in (Chen et al., 2001).

Consider a control function which acts on the FODE system (3.8) as follows:

$$a_n D_t^{\alpha_n} y(t) + \cdots + a_1 D_t^{\alpha_1} y(t) + a_0 D_t^{\alpha_0} y(t) = u(t). \tag{3.13}$$

By Laplace transform, we can get a fractional transfer function:

$$G(s) = \frac{Y(s)}{U(s)} = \frac{1}{a_n s^{\alpha_n} + \cdots + a_1 s^{\alpha_1} + a_0 s^{\alpha_0}}. \tag{3.14}$$

The fractional-order linear time-invariant (LTI) system can also be represented by the following state-space model (Matignon, 1998):

$$\begin{aligned} {}_0D_t^{\mathbf{q}} x(t) &= \mathbf{A}x(t) + \mathbf{B}u(t), \\ y(t) &= \mathbf{C}x(t), \end{aligned} \tag{3.15}$$

where $x \in \mathrm{R}^n$, $u \in \mathrm{R}^r$ and $y \in \mathrm{R}^p$ are the state, input and output vectors of the system and $\mathbf{A} \in \mathrm{R}^{n\times n}$, $\mathbf{B} \in \mathrm{R}^{n\times r}$, $\mathbf{C} \in \mathrm{R}^{p\times n}$, and $\mathbf{q} = [q_1, q_2, \ldots, q_n]^T$ are the fractional orders. If $q_1 = q_2 = \cdots = q_n \equiv \alpha$, system (3.15) is called a commensurate-order system, otherwise it is an incommensurate-order system.

The state transition matrix is

$$\mathbf{x}(t) = \left[\mathbf{I}+\frac{\mathbf{A}\mathbf{x}(0)}{\Gamma(1+\alpha)}t^{\alpha}+\frac{\mathbf{A}^2\mathbf{x}(0)}{\Gamma(1+2\alpha)}t^{2\alpha}+\cdots+\frac{\mathbf{A}^k\mathbf{x}(0)}{\Gamma(1+k\alpha)}t^{k\alpha}+\cdots\right]$$
$$= \left(\sum_{k=0}^{\infty}\frac{\mathbf{A}^k t^{k\alpha}}{\Gamma(1+k\alpha)}\right)\mathbf{x}(0) = \phi(\mathbf{t})\mathbf{x}(\mathbf{0}). \tag{3.16}$$

Similar to conventional observability and controllability concept, the controllability is defined as follows (Matignon and D'Andrea-Novel, 1996): System (3.15) is *controllable* on $[t_0, t_{final}]$ if the controllability matrix

$$C_a = [B|AB|A^2B|\cdots|A^{n-1}B]$$

has rank n.

The observability is defined as follows (Matignon and D'Andrea-Novel, 1996): System (3.15) is *observable* on $[t_0, t_{final}]$ if the observability matrix

$$O_a = \begin{bmatrix} C \\ CA \\ CA^2 \\ \vdots \\ CA^{n-1} \end{bmatrix}$$

has rank n.

A fractional-order system described by n-term fractional differential equation (3.13) can be rewritten into the state-space representation in the form (Dorčák et al., 2002; Yang and Liu, 2006):

$$\begin{bmatrix} {}_0D^{q_1}x_1(t) \\ {}_0D^{q_2}x_2(t) \\ \vdots \\ {}_0D^{q_n}x_n(t) \end{bmatrix} = \begin{bmatrix} 0 & 1 & \cdots & 0 \\ 0 & 0 & 1 & 0 \\ \vdots & \vdots & & \vdots \\ -a_0/a_n & -a_1/a_n & \cdots & a_{n-1}/a_n \end{bmatrix} \begin{bmatrix} x_1(t) \\ x_2(t) \\ \vdots \\ x_n(t) \end{bmatrix} + \begin{bmatrix} 0 \\ 0 \\ \vdots \\ 1/a_n \end{bmatrix} u(t)$$
$$y(t) = \begin{bmatrix} 1 \; 0 \cdots 0 \; 0 \end{bmatrix} \begin{bmatrix} x_1(t) \\ x_2(t) \\ \vdots \\ x_n(t) \end{bmatrix}, \tag{3.17}$$

where $\alpha_0 = 0$, $q_1 = \alpha_1$, $q_2 = \alpha_{n-1} - \alpha_{n-2}, \ldots, q_n = \alpha_n - \alpha_{n-1}$, and with initial conditions:

$$x_1(0) = x_0^{(1)} = y_0, \; x_2(0) = x_0^{(2)} = 0, \ldots$$

$$x_i(0) = x_0^{(i)} = \begin{cases} y_0^{(k)}, & \text{if } i = 2k+1, \\ 0, & \text{if } i = 2k, \end{cases} \quad i \leq n. \tag{3.18}$$

The n-term FODE (3.13) is equivalent to the system of Eqs. (3.17) with the initial conditions (3.18) if Caputo's derivative is considered.

3.2 Fractional Nonlinear Systems

In this book, we will consider the general incommensurate fractional-order nonlinear system represented as follows:

$$\begin{aligned} {}_0D_t^{q_i} x_i(t) &= f_i(x_1(t), x_2(t), \ldots, x_n(t), t), \\ x_i(0) &= c_i, \; i = 1, 2, \ldots, n, \end{aligned} \tag{3.19}$$

where c_i are initial conditions. The vector representation of (3.19) is:

$$D^{\mathbf{q}}\mathbf{x} = \mathbf{f}(\mathbf{x}), \tag{3.20}$$

where $\mathbf{q} = [q_1, q_2, \ldots, q_n]^T$ for $0 < q_i < 2$, $(i = 1, 2, \ldots, n)$ and $\mathbf{x} \in \mathrm{R}^n$.

The equilibrium points of system (3.20) are calculated via solving the following equation

$$\mathbf{f}(\mathbf{x}) = 0 \tag{3.21}$$

and we suppose that $E^* = (x_1^*, x_2^*, \ldots, x_n^*)$ is an equilibrium point of system (3.20).

3.3 Fractional-Order Controllers

3.3.1 Definition of Fractional-Order Controllers

The fractional-order $PI^{\lambda}D^{\delta}$ (also $PI^{\lambda}D^{\mu}$ controller) controller (FOC) was proposed in (Podlubny, 1999a,b) as a generalization of the PID controller with integrator of real order λ and differentiator of real order δ. The transfer function of such controller in the Laplace domain has this form (Podlubny, 1999b; Podlubny et al., 2002):

$$C(s) = \frac{U(s)}{E(s)} = K_p + T_i s^{-\lambda} + T_d s^{\delta}, \qquad (\lambda, \delta > 0), \tag{3.22}$$

where K_p is the proportional constant, T_i is the integration constant and T_d is the differentiation constant.

As we can see in Fig. 3.1, the internal structure of the fractional-order controller consists of the parallel connection, the proportional, integration, and derivative part (Dorf and Bishop, 1990). The transfer function (3.22) corresponds in time domain

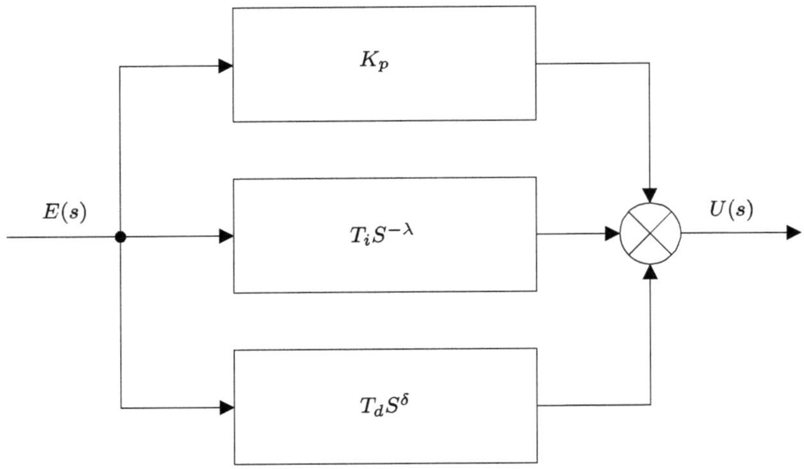

Fig. 3.1 General structure of a $PI^{\lambda}D^{\delta}$ controller.

to fractional differential equation (3.23)

$$u(t) = K_p e(t) + T_i\, {}_0D_t^{-\lambda} e(t) + T_d\, {}_0D_t^{\delta} e(t), \tag{3.23}$$

or discrete transfer function given below:

$$C(z) = \frac{U(z)}{E(z)} = K_p + \frac{T_i}{(\omega(z^{-1}))^{\lambda}} + T_d(\omega(z^{-1}))^{\delta}, \tag{3.24}$$

where $\omega(z^{-1})$ denotes the discrete operator, expressed as a function of the complex variable z or the shift operator z^{-1}.

Taking $\lambda = 1$ and $\delta = 1$, we obtain a classical *PID* controller. If $\lambda = 0$ and $T_i = 0$, we obtain a PD^{δ} controller, etc. All these types of controllers are particular cases of the fractional-order controller, which is more flexible and gives an opportunity to better adjust the dynamical properties of the fractional-order control system.

It can also be mentioned that there are many other considerations of the fractional-order controller (Xue and Chen, 2002). For example, we can mention several of them:

- *CRONE* controller (1st generation) (Oustaloup, 1995), characterized by the band-limited lead effect (Oustaloup, 1983):

$$C(s) = C_0 \left(\frac{1 + s/\omega_b}{1 + s/\omega_h} \right)^r \tag{3.25}$$

  where $0 < \omega_b < \omega_h$, $C_0 > 0$ and $r \in (0,1)$.
  There are a number of real life applications of three generations of the CRONE controller such as the car suspension control (Oustaloup et al., 1996), flexible transmission (Oustaloup et al., 1995), and hydraulic actuator.

- Fractional lead-lag compensator (Raynaud and Zergaïnoh, 2000; Monje et al., 2008), given by

$$C(s) = k_c \left(\frac{s+1/\lambda}{s+1/x\lambda} \right)^r = k_c x^r \left(\frac{\lambda s+1}{x\lambda s+1} \right)^r, \tag{3.26}$$

$$r \in \mathrm{R},\ 0 < x < 1.$$

- Non-integer integral and its application to control (Manabe, 1961);
- *TID* compensator (Lurie, 1994), which has structure similar to a *PID* controller but the proportional component is replaced with a tilted component having a transfer function s to the power of $(-1/n)$. The resulting transfer function of the *TID* controller has the form:

$$C(s) = \frac{T}{s^{1/n}} + \frac{I}{s} + Ds, \tag{3.27}$$

where T, I and D are the controller constants and n is a non-zero real number, preferably between 2 and 3. The transfer function (3.27) more closely approximates an optimal transfer function and an overall response is achieved, which is closer to the theoretical optimal response determined by Bode (Bode, 1949).

3.3.2 Properties and Characteristics of Controller

It can be expected that $PI^\lambda D^\delta$ controller (3.22) may enhance the systems control performance due to more tuning knobs introduced, which is intuitively illustrated in Fig. 3.2.

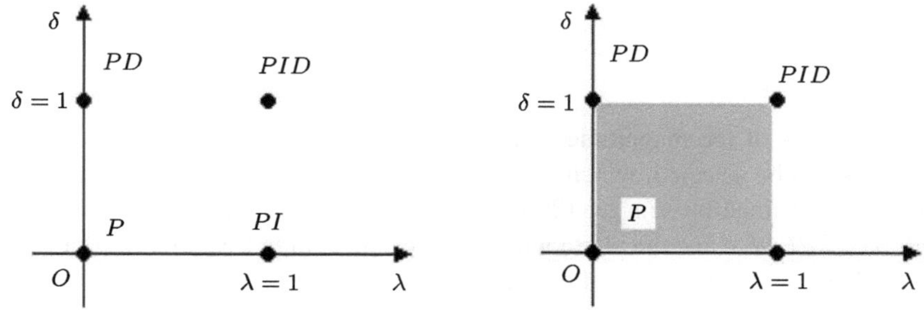

Fig. 3.2 *PID* controller: from *points* to *plane*.

The $PI^\lambda D^\delta$ controller with complex zeros and poles located anywhere in the left-hand s-plane may be rewritten as

$$C(s) = K\frac{(s/\omega_n)^{\delta+\lambda} + (2\zeta s^\lambda)/\omega_n + 1}{s^\lambda}, \tag{3.28}$$

where K is the gain, ζ is the dimensionless damping ratio and ω_n is the natural frequency. Normally, we choose $\zeta < 1$. When $\zeta = 1$, the condition is called critical damping (Dorf and Bishop, 1990).

Example 3.1. : Let us consider the fractional-order controller (3.28) with the following parameters: $K = 6.5$, $\omega_n = 1$, $\zeta = 0.5$ and $\lambda = \delta = 0.5$.

In Fig. 3.3 are shown the Bode plots of the fractional $PI^\lambda D^\delta$ controller (3.28) with the above-mentioned parameters.

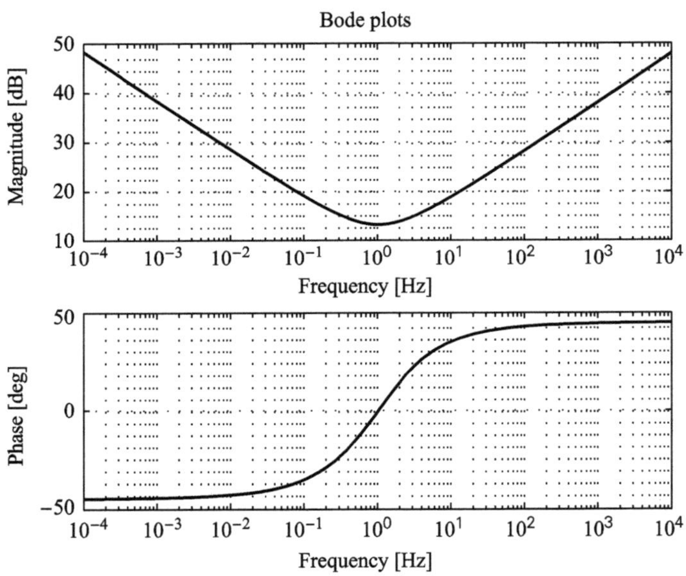

Fig. 3.3 Bode plots of $PI^\lambda D^\delta$ controller (3.28) with $K = 6.5$, $\omega_n = 1$, $\zeta = 0.5$ and $\lambda = \delta = 0.5$.

The slopes of the magnitude and the value of the phase for low and high frequencies can be selected, which are related with the relative stability and the high frequencies gained by Vinagre (2000). Asymptotically, at low frequency the slope will be $-\lambda dB/dec$ and the phase will be $-\lambda\frac{\pi}{2}$, and at high frequency the slope will be $\delta dB/dec$ and the phase will be $\delta\frac{\pi}{2}$.

For a wide class of controlled objects we recommend the fractional PI^nD^δ controller, which is a particular case of $PI^\lambda D^\delta$ controller, where $\lambda = n$, $n \in N$ and $\delta \in R$. Integer-order integrator is important for steady-state error cancellation but on the other hand, the fractional integral is also important for obtaining a *Bode's ideal loop transfer function* response with constant phase margin for desired frequency range (Aström, 2000; Bode, 1949; Manabe, 1961; Tustin et al., 1958).

3.3.3 Design of Controller Parameters and Implementation

The tuning of $PI^{\lambda}D^{\delta}$ controller parameters is determined according to the given requirements. These requirements are, for example, the damping ratio, the steady-state error (e_{ss}), dynamical properties, etc. One of the methods being developed is the method of dominant roots (Petráš, 1999; Petráš and Dorčák, 2003), based on the given stability measure and the damping ratio of the closed control loop. Assume that the desired dominant roots are a pair of complex conjugate root as follows:

$$s_{1,2} = -\sigma \pm j\omega_d,$$

designed for the damping ratio ζ and natural frequency ω_n. The damping constant (stability measure) is $\sigma = \zeta\omega_n$ and the damped natural frequency of oscillation $\omega_d = \omega_n\sqrt{1-\zeta^2}$. The design of parameters: K_p, T_i, λ, T_d and δ can be computed numerically from characteristic equation. More specifically, for simple plant model $P(s)$, this can be done by solving

$$\min_{K_p,T_i,\lambda,T_d,\delta} ||C(s)P(s)+1||_{s=-\sigma\pm j\omega_d}.$$

Another possible way to obtain the controller parameters is using the tuning formula, based on gain A_m and phase Φ_m margins specifications for crossover frequency ω_{cg}. Gain and phase margins have always served as important measures of robustness. The equations that define the phase margin and the gain crossover frequency are expressed as (Monje et al., 2008; Vinagre, 2000):

$$|C(j\omega_{cg})P(\omega_{cg})|_{\text{dB}} = 0\,\text{dB}$$

$$\arg(C(j\omega_{cg})P(\omega_{cg})) = -\pi + \Phi_m$$

The above equations are often used also for the so-called auto-tuning techniques. For instance, relay auto-tuning process has been widely used in industrial application and it was already modified for the fractional-order controllers (Monje et al., 2008).

Last but not least, we should mention the optimization algorithm based on the integral absolute error (IAE) minimization (Podlubny, 1999b):

$$\text{IAE}(t) = \int_0^t |e(t)|dt = \int_0^t |w(t)-y(t)|dt,$$

where $w(t)$ is the desired value of closed control loop and $y(t)$ is the real value of closed control loop. This method does not ensure the desired stability measure of the closed control loop. Measure of stability has to be checked out additionally by some known method as, for example, frequency method described in literature (Petráš and Dorčák, 1999).

Implementation techniques for the FOC have been described in several works. Some proposal can be found in the work by Vinagre (2000). An analogue implementation was proposed in the book by Petráš et al. (2002) and a digital implementation was suggested in the book by Caponetto et al. (2010).

It can be expected that $PI^{\lambda}D^{\delta}$ controller (3.22) may enhance the systems control performance due to more tuning knobs introduced. Actually, in theory, $PI^{\lambda}D^{\delta}$ itself is an infinite-dimensional linear filter due to the fractional order in differentiator or integrator.

We comment that since *PID* control is ubiquitous in industry process control, fractional-order *PID* control will also be ubiquitous when tuning and implementation techniques are well developed (Caponetto et al., 2010; Chen, 2006; Chen et al., 2008, 2009; Monje et al., 2008; Petráš, 1999).

References

Aström K. J., 2000, *Model Uncertainty and Robust Control*, COSY project.

Bayat F. M. and Afshar M., 2008, Extending the root-locus method to fractional-order systems, *Journal of Applied Mathematics*, Article ID **528934**.

Bode H. W., 1949, *Network Analysis and Feedback Amplifier Design*, Tung Hwa Book Company, Shanghai.

Caponetto R., Dongola G., Fortuna L. and Petráš I., 2010, *Fractional Order Systems: Modeling and Control Applications*, World Scientific, Singapore.

Chen Y. Q., 2006, Ubiquitous Fractional Order Controls? *Proc. of the Second IFAC Symposium on Fractional Derivatives and Applications (IFAC FDA06)*, July 19–21, Porto, Portugal.

Chen Y. Q., Petráš I. and Vinagre B. M., 2001, A List of Laplace and Inverse Laplace Transforms Related to Fractional Order Calculus, http://people.tuke.sk/ivo.petras/foc_laplace.pdf.

Chen Y. Q., Bhaskaran T. and Xue D., 2008, Practical tuning rule development for fractional order proportional and integral controllers, *ASME Journal of Computational and Nonlinear Dynamics*, **3**, 021403-1 – 021403-8.

Chen Y. Q., Petráš I. and Xue D., 2009, Fractional order control – A tutorial, *Proc. of the American Control Conference, ACC 2009.*, June 10–12, 2009, St. Louis, USA, 1397–1411.

Dorčák Ľ, Petráš I., Koštial I. and Terpák J., 2002, Fractional-order state space models, *Proc. of the International Carpathian Control Conference*, Malenovice, Czech republic, May 27-30, 193–198.

Dorf R. C. and Bishop R. H., 1990, *Modern Control Systems*, Addison-Wesley, New York.

Gorenflo R., Luchko Yu. and Rogosin S., 2004, Mittag-Leffler type functions: notes on growth properties and distribution of zeros, Preprint No. **A-97-04**, Fachbereich Mathematik und Informatik, Freie Universitat Berlin, Germany.

LePage W. R., 1961, *Complex Variables and the Laplace Transform for Engineers*, McGraw-Hill, New York.

Lurie B. J., 1994, Three-Parameter Tunable Tilt-Integral-Derivative (TID) Controller, *United States Patent, 5 371 670*, USA.

Magin R. L., Feng X. and Baleanu D., 2009, Solving the fractional order Bloch equation, *Concepts in Magnetic Resonance Part A*, **34A**, 16–23.

Manabe S., 1961, The non-integer integral and its application to control systems, *ETJ of Japan*, **6**, 83–87.

Matignon D., 1998, Generalized fractional differential and difference equations: stability properties and modelling issues, *Proc. of the Math. Theory of Networks and Systems Symposium*, Padova, Italy.

Matignon D. and D'Andrea-Novel B., 1996, Some results on controllability and observability of finite-dimensional fractional differential systems, *Computational Engineering in Systems Applications*, Lille, France, IMACS, IEEE-SMC, **2**, 952–956.

Monje C. A., Vinagre B. M., Feliu V. and Chen Y. Q., 2008, Tuning and auto-tuning of fractional order controllers for industry application, *Contr. Eng. Pract.*, **16**, 798–812.

Oustaloup A., 1983, *Systèmes Asservis Linéaires d'Ordre Fractionnaire: Théorie et Pratique*, Editions Masson, Paris.

Oustaloup A., 1995, *La Derivation Non Entiere: Theorie, Synthese et Applications*, Hermes, Paris.

Oustaloup A., Mathieu B. and Lanusse P., 1995, The CRONE control of resonant plants: application to a flexible transmission, *European Journal of Control*, **1**, 113–121.

Oustaloup A., Moreau X. and Nouillant M., 1996, The CRONE suspension, *Control Engineering Practice*, **4**, 1101–1108.

Petráš I., 1999, The fractional-order controllers: methods for their synthesis and application, *Journal of Electrical Engineering*, **50**, 284–288.

Petráš I. and Dorčák Ľ., 1999, The frequency method for stability investigation of fractional control systems, *Journal of SACTA*, **2**, 75–85.

Petráš I., Dorčák Ľ., O'Leary P., Vinagre B. M. and Podlubny I., 2000, The modelling and analysis of fractional-order control systems in frequency domain, *Proc. of the ICCC'2000 Conference*, May 23-26, High Tatras, 261–264.

Petráš I., Podlubny I., O'Leary P., Dorčák Ľ. and Vinagre B. M., 2002, *Analogue Realization of Fractional Order Controllers*, Faculty of BERG, Technical University of Kosice.

Petráš I. and Dorčák Ľ., 2003, Fractional-order control systems: Modelling and simulation, *Fractional Calculus and Applied Analysis*, **6**, 205–232.

Podlubny I., 1999a, *Fractional Differential Equations*, Academic Press, San Diego.

Podlubny I., 1999b, Fractional-order systems and $PI^{\lambda}D^{\mu}$-controllers, *IEEE Transactions on Automatic Control*, **44**, 208–213.

Podlubny I., Petráš, I., Vinagre, B.M., O'Leary P. and Dorčák Ľ., 2002, Analogue realizations of fractional-order controllers, *Nonlinear Dynamics*, **29**, 281–296.

Raynaud H. F. and Zergaïnoh A., 2000, State-space representation for fractional order controllers, *Automatica*, **36**, 1017–1021.

Tustin A., Allanson J. T., Layton J. M. and Jakeways R. J., 1958, The design of systems for automatic control of the position of massive objects, *The Proceedings of the Institution of Electrical Engineers*, **105C**.

Vinagre B. M., Podlubny I., Hernandez A., and Feliu V., 2000, On realization of fractional-order controllers, *Proc. of the Conference Internationale Francophone d'Automatique*, Lille, Jule 5-8, 945–950.

Xue D. and Chen Y. Q., 2002, A comparative introduction of four fractional order controllers, *Proc. of the* 4th *World Congress on Intelligent Control and Automation*, June 10 - 14, Shanghai, China.

Yang C. and Liu F., 2006, A computationally effective predictor-corrector method for simulating fractional order dynamical control system, *Australian and New Zealand Industrial and Applied Mathematics Journal*, **47**, C168–C184.

Chapter 4
Stability of Fractional-Order Systems

4.1 Preliminary Consideration

Stability as an extremely important property of dynamical systems can be investigated in various domains (Bellman, 1953; Dorf and Bishop, 1990). The usual concept of the bounded input-bounded output (BIBO) or external stability in *time domain* can be defined via the following general stability conditions (Matignon, 1998):

A causal LTI system with impulse response $h(t)$ will be BIBO stable if the necessary and sufficient condition is satisfied

$$\int_0^\infty ||h(\tau)||d\tau < \infty,$$

where the output of the system is defined by convolution

$$y(t) = h(t) * u(t) = \int_0^\infty h(\tau)u(t-\tau)d\tau,$$

where $u, y \in L_\infty$ and $h \in L_1$.

Another very important domain is *frequency domain*. In the case of frequency method for evaluating the stability we transform the s-plane into the complex plane $G_o(j\omega)$ and the transformation is realized according to the transfer function of the open loop system $G_o(j\omega)$. During the transformation, all roots of the characteristic polynomial are mapped from s-plane into the critical point $(-1, j0)$ in the plane $G_o(j\omega)$. The mapping of the s-plane into $G_o(j\omega)$ plane is conformal, that is, the direction and location of points in the s-plane are preserved in the $G_o(j\omega)$ plane. Frequency investigation method and utilization of the Nyquist frequency characteristics based on argument principle were described in the paper by Petráš and Dorčák (1999).

However, we cannot directly use algebraic tools as, for example, Routh-Hurwitz criteria for the fractional-order system because we do not have a characteristic polynomial but pseudo-polynomial with rational power — *multivalued function*. It is possible only in some special cases (Ahmed et al., 2006). Moreover, modern control methods as, for example, LMI (Linear Matrix Inequality) methods (Oustaloup et al., 2008) or other algorithms (Hamamci, 2008; Hwang and Cheng, 2006) have already been developed. The advantage of LMI methods in control theory is due to their connection with the Lyapunov method (existence of a quadratic Lyapunov function). More generally, LMI methods are useful to test if matrix eigenvalues belong to a certain region in the complex plane. A simple test can be used (Anderson et al., 1974). Roots of polynomial $P(s) = \det(sI - A)$ lie inside the region $-\pi/2 - \delta < \arg(s) < \pi/2 + \delta$ if eigenvalues of the matrix

$$A_1 = \begin{bmatrix} A\cos\delta & -A\sin\delta \\ A\sin\delta & A\cos\delta \end{bmatrix} \equiv A \otimes \begin{bmatrix} \cos\delta & -\sin\delta \\ \sin\delta & \cos\delta \end{bmatrix} \tag{4.1}$$

have negative real part, where $\otimes$ denotes the Kronecker product. This property has been used in stability analysis of ordinary fractional order LTI systems and also for interval fractional-order LTI systems (Tavazoei and Haeri, 2009).

The stability of the fractional-order delayed systems was investigated in (Bonnet and Partington, 2000; Chen and Moore, 2002; Özturk and Uraz, 1985) as well as the stability of the discrete fractional-order systems in (Dzieliski and Sierociuk, 2008).

When dealing with incommensurate fractional-order systems (or, in general, with fractional-order systems), it is important to bear in mind that $P(s^\alpha)$, $\alpha \in R$ is a multivalued function of s^α, $\alpha = u/v$, the domain of which can be viewed as a Riemann surface with finite number of Riemann sheets v, where the origin is a branch point and the branch cut is assumed at R^- (Fig. 4.1). Function s^α becomes holomorphic in the complement of the branch cut line. It is a fact that in multivalued functions only the first Riemann sheet has its physical significance (Gross and Braga, 1961). Note that each Riemann sheet has only one edge at branch cut and not only poles and singularities originating from the characteristic equation, but branch points and branch cut of given multivalued functions are also important for the stability analysis (Bayat and Afshar, 2008; Bayat and Ghartemani, 2008).

In this book the branch cut is assumed at R^- and the first Riemann sheet is denoted by Ω and defined as (Fig. 4.1)

$$\Omega := \{re^{j\phi} \,|\, r > 0, -\pi < \phi < \pi\}. \tag{4.2}$$

It is well known that an integer-order LTI system is stable if all the roots of the characteristic polynomial $P(s)$ are negative or have negative real parts if they are complex conjugate (e.g. (Dorf and Bishop, 1990)). This means that they are located on the left of the imaginary axis of the complex s-plane. System $G(s) = Q(s)/P(s)$ is BIBO stable if

$$\exists \;\; ||G(s)|| \le M < \infty, \;\; M > 0, \;\; \forall s, \Re(s) \ge 0.$$

A necessary and sufficient condition for the asymptotic stability is (El-Salam and El-Sayed, 2007):

$$\lim_{t\to\infty} ||X(t)|| = 0.$$

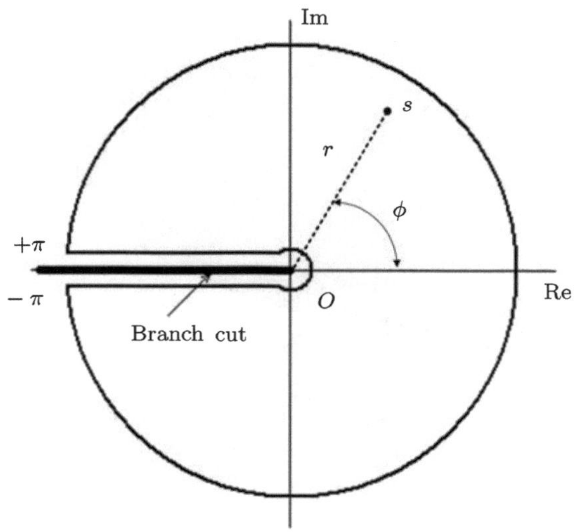

Fig. 4.1 Branch cut $(0, -\infty)$ for branch points in the complex plane.

According to the final value theorem proposed in (Ghartemani and Bayat, 2008), for fractional-order case, when there is a branch point at $s = 0$, we assume that $G(s)$ is a multivalued function of s, then

$$x(\infty) = \lim_{s \to 0}[sG(s)].$$

Example 4.1. : Let us investigate the simplest multivalued function defined as

$$w = s^{\frac{1}{2}} \tag{4.3}$$

and there will be two s-planes which map onto a single w-plane. The interpretation of the two sheets of the Riemann surface and the branch cut is depicted in Fig. 4.2.

Define the principal square root function as

$$f_1(s) = |s|^{\frac{1}{2}} e^{\frac{j\phi}{2}} = re^{\frac{j\phi}{2}},$$

where $r > 0$ and $-\pi < \phi < +\pi$. The function $f_1(s)$ is a branch of w. Using the same notation, we can find other branches of the square root function. For example, if we let

$$f_2(s) = |s|^{\frac{1}{2}} e^{\frac{j\phi + 2\pi}{2}} = re^{\frac{j\phi + 2\pi}{2}},$$

then $f_2(s) = -f_1(s)$ and it can be thought of as "plus" and "minus" square root functions. The negative real axis is called a branch cut for the functions $f_1(s)$ and $f_2(s)$. Each point on the branch cut is a point of discontinuity for both functions $f_1(s)$ and $f_2(s)$. As has been shown in (LePage, 1961), the function described by

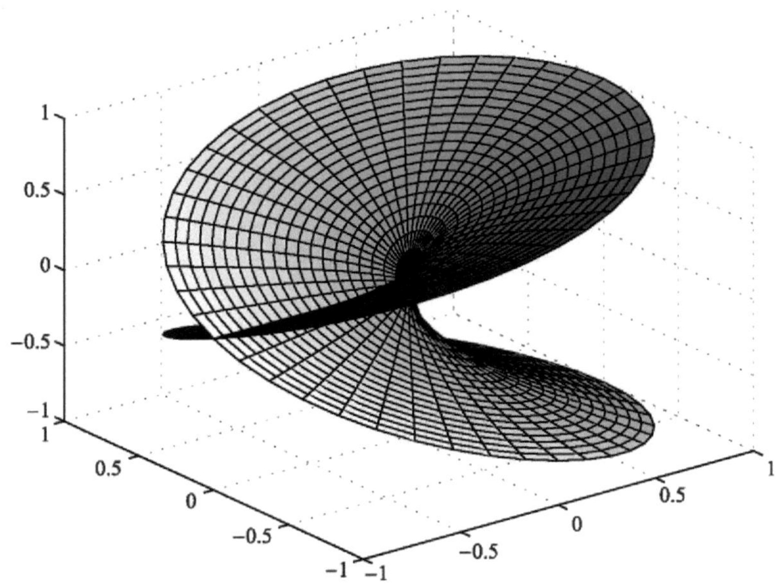

Fig. 4.2 Riemann surface interpretation of the function $w = s^{\frac{1}{2}}$.

(4.3) has a branch point of order 1 at $s = 0$ and at infinity. They are located at the ends of the branch cut (Fig. 4.1).

Example 4.2. : Let us investigate the transfer function of fractional-order system (multivalued function) defined as

$$G(s) = \frac{1}{s^{\alpha} + b}, \tag{4.4}$$

where $\alpha \in \mathrm{R}$ $(0 < \alpha \leq 2)$ and $b \in \mathrm{R}$ $(b > 0)$.

The analytical solution of the fractional-order system (4.4) obtained according to relation (3.9) has the following form:

$$g(t) = \mathscr{E}_0(t, -b; \alpha, \alpha). \tag{4.5}$$

The Riemann surface of the function (4.4) contains an infinite number of sheets and infinitely many poles in positions

$$s = b^{\frac{1}{\alpha}} e^{\frac{j(\pi + 2\pi n)}{\alpha}}, \quad n = 0, \pm 1, \pm 2, \ldots \text{ for } \alpha > 0 \, \text{and} \, b > 0.$$

The sheets of the Riemann surface are all different if α is irrational.

For $1 < \alpha < 2$ we have two poles corresponding to $n = 0$ and $n = -1$, and the poles are

$$s = b^{\frac{1}{\alpha}} e^{\pm \frac{i\pi}{\alpha}}.$$

However, for $0 < \alpha < 1$ in (4.4) the denominator is a multivalued function and the singularity of the system cannot be defined unless it is made singlevalued. Therefore we will use the Riemann surface. Let us investigate transfer function (4.4) for $\alpha = 0.5$ (half-order system), then we get

$$G(s) = \frac{1}{s^{\frac{1}{2}} + b}, \tag{4.6}$$

and by equating the denominator to zero we have

$$s^{\frac{1}{2}} + b = 0.$$

Rewriting the complex operator $s^{\frac{1}{2}}$ in exponential form and using the well-known relation $e^{j\pi} + 1 = 0$ (or $e^{j(\pm\pi + 2k\pi)} + 1 = 0$) we get the following formula:

$$r^{\frac{1}{2}} e^{j(\phi/2 + k\pi)} = a e^{j(\pm\pi + 2k\pi)} \tag{4.7}$$

From relationship (4.7) it can be deduced that the modulus and phase (arg) of the pole are:

$$r = b^2 \text{ and } \phi = \pm 2\pi(1+k) \text{ for } k = 0, 1, 2, \ldots$$

However, the first sheet of the Riemann surface is defined for range of $-\pi < \phi < +\pi$, the pole with the angle $\phi = \pm 2\pi$, does not fall within this range but the pole with the angle $\phi = 2\pi$ falls into the range of the second sheet defined for $\pi < \phi < 2\pi$. Therefore this half-order pole with magnitude b^2 is located on the second sheet of the Riemann surface that consequently maps to the left side of the w-plane (Fig. 4.3). On this plane the magnitude and phase of the singlevalued pole are b^2 and π, respectively (LePage, 1961).

Example 4.3. : Analogous to previous examples we can also investigate function

$$w = s^{\frac{1}{3}}, \tag{4.8}$$

where in this case the Riemann surface has three sheets with each mapping onto one third of the w-plane (Fig. 4.4).

Definition 4.1. Generally, for the multivalued function defined as

$$w = s^{\frac{1}{v}}, \tag{4.9}$$

where $v \in \mathrm{N}$ ($v = 1, 2, 3, \ldots$) we get the v sheets in the Riemann surface. In Fig. 4.5 is shown the relationship between the w-plane and the v sheets of the Riemann surface where sector $-\pi/v < \arg(w) \leq \pi/v$ corresponds to Ω (first Riemann sheet).

Definition 4.2. Mapping the poles from the s^q-plane into the w-plane, where $q \in \mathrm{Q}$ is such that $q = \dfrac{k}{m}$ for $k, m \in \mathrm{N}$ and $|\arg(w)| = |\phi|$, can be done by the following

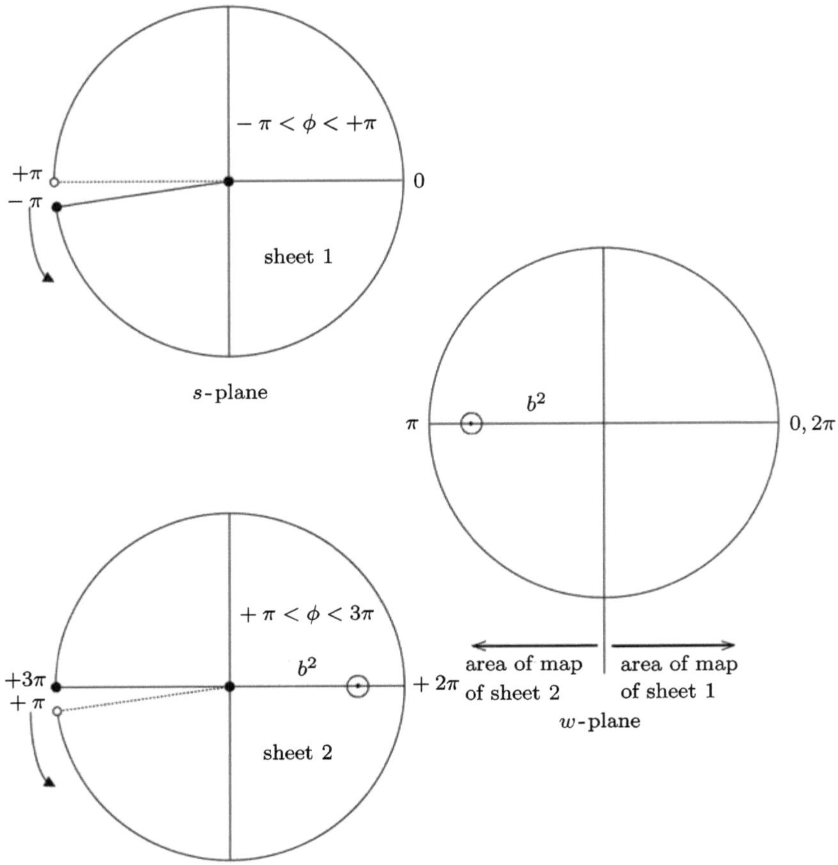

Fig. 4.3 Correspondence between the s-plane and the w-plane.

rule: If we assume $k = 1$, then the mapping from s-plane to w-plane is independent of k. Unstable region from the s-plane transforms to sector $|\phi| < \frac{\pi}{2m}$ and stable region transforms to sector $\frac{\pi}{2m} < |\phi| < \frac{\pi}{m}$. The region where $|\phi| > \frac{\pi}{m}$ is not physical. Therefore, the system will be stable if all roots in the w-plane lie in the region $|\phi| > \frac{\pi}{2m}$. Stability regions depicted in Fig. 4.6 correspond to the following propositions:

1. For $k < m$ $(q < 1)$ the stability region is depicted in Fig. 4.6(a).
2. For $k = m$ $(q = 1)$ the stability region corresponds to the s-plane.
3. For $k > m$ $(q > 1)$ the stability region is depicted in Fig. 4.6(b).

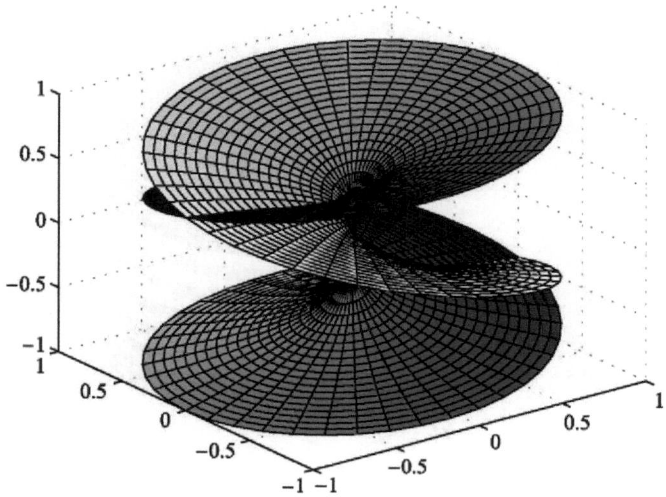

(a) Riemann surface

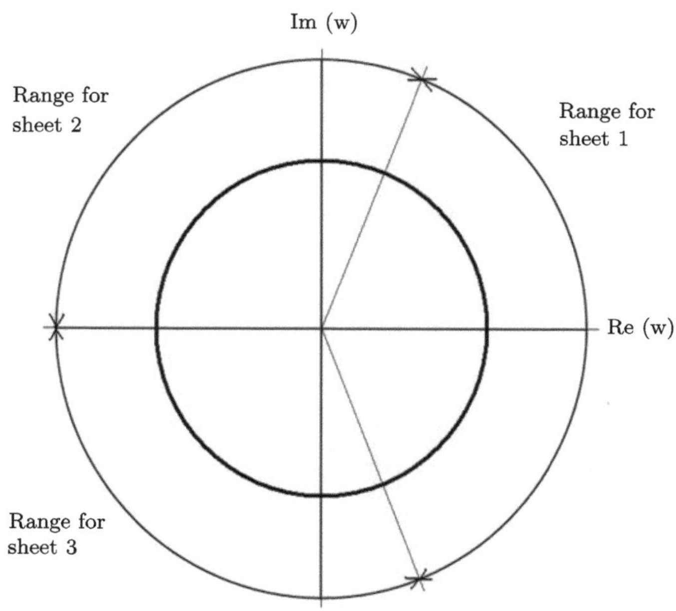

(b) Complex w-plane

Fig. 4.4 Correspondence between the 3-sheet Riemann surface and w-plane for Eq. (4.8).

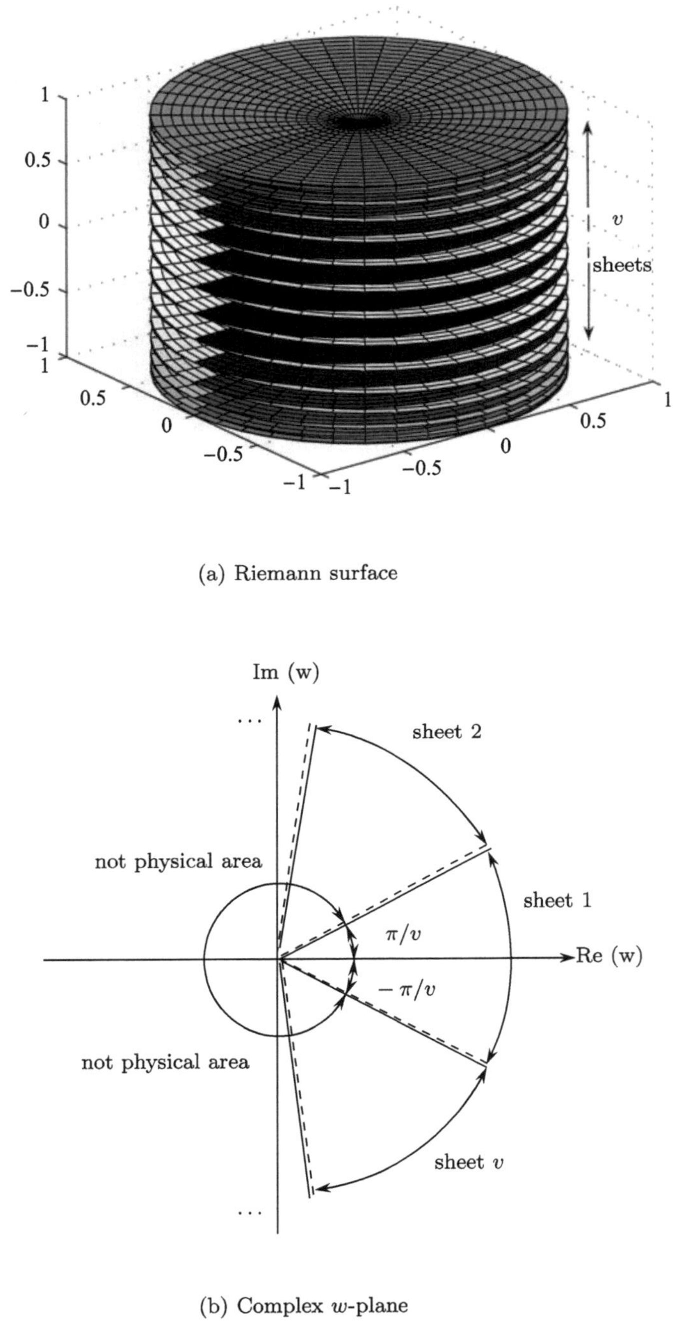

Fig. 4.5 Correspondence between the w-plane and the Riemann sheets.

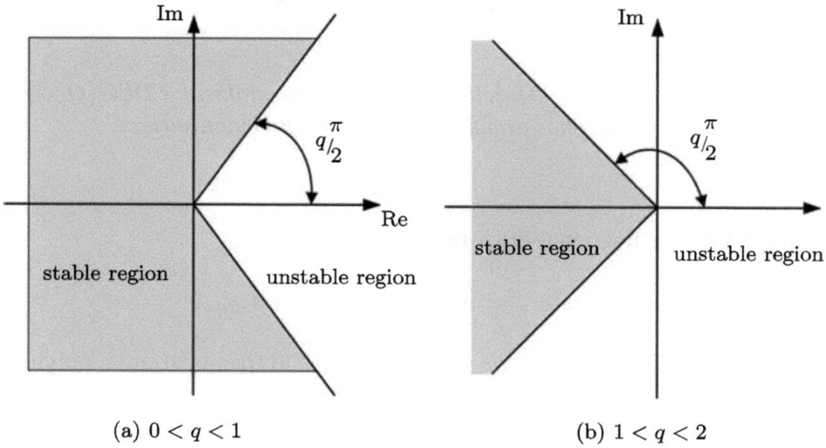

Fig. 4.6 Stability regions of the fractional-order system.

4.2 Stability of Fractional LTI Systems

As we can see in previous subsection, in the fractional case, the stability is different from the integer case. It is interesting that a stable fractional system may have roots in the right half of complex w-plane (Fig. 4.6). Since the principal sheet of the Riemann surface is defined for $-\pi < \arg(s) < \pi$, by using the mapping $w = s^q$, the corresponding w domain is defined by $-q\pi < \arg(w) < q\pi$, and the w plane region corresponding to the right half plane of this sheet is defined by $-q\pi/2 < \arg(w) < q\pi/2$.

Consider the fractional-order pseudo-polynomial

$$Q(s) = a_1 s^{q_1} + a_2 s^{q_2} + \cdots + a_n s^{q_n} = a_1 s^{c_1/d_1} + a_2 s^{c_2/d_2} + \cdots + a_n s^{c_n/d_n},$$

where q_i are rational numbers expressed as c_i/d_i and a_i are real numbers for $i = 1, 2, \ldots, n$. If for some i, $c_i = 0$ then $d_i = 1$. Let v be the least common multiple (LCM) of $d_1, d_2, \ldots, d_n$ denoted as $v = \mathrm{LCM}\{d_1, d_2, \ldots, d_n\}$, then (Ghartemani and Bayat, 2008)

$$Q(s) = a_1 s^{\frac{v_1}{v}} + a_2 s^{\frac{v_2}{v}} + \cdots + a_n s^{\frac{v_n}{v}} = a_1 (s^{\frac{1}{v}})^{v_1} + a_2 (s^{\frac{1}{v}})^{v_2} + \cdots + a_n (s^{\frac{1}{v}})^{v_n}. \quad (4.10)$$

The fractional degree (FDEG) of the polynomial $Q(s)$ is defined as (Ghartemani and Bayat, 2008)

$$\mathrm{FDEG}\{Q(s)\} = \max\{v_1, v_2, \ldots, v_n\}.$$

The domain of definition for (4.10) is the Riemann surface with v Riemann sheets where the origin is a branch point of order $v - 1$ and the branch cut is assumed at R^-. The number of roots for fractional algebraic equation (4.10) is given by the following proposition (Bayat et al., 2009):

Proposition 4.1. *Let $Q(s)$ be a fractional-order polynomial with FDEG$\{Q(s)\} = n$. Then the equation $Q(s)=0$ has exactly n roots on the Riemann surface.*

Definition 4.3. The fractional-order polynomial

$$Q(s) = a_1 s^{\frac{n}{v}} + a_2 s^{\frac{n-1}{v}} + \cdots + a_n s^{\frac{1}{v}} + a_{n+1}$$

is *minimal* if FDEG$\{Q(s)\} = n$. We will assume that all fractional-order polynomial are minimal. This ensures that there is no redundancy in the number of the Riemann sheets (Ghartemani and Bayat, 2008).

On the other hand, it has been shown, by several authors and by using several methods, that for the case of FOLTI system of commensurate order, a geometrical method of complex analysis based on the argument principle of the roots of the characteristic equation (a polynomial in this particular case) can be used in the stability check in the BIBO sense (Matignon, 1998; Petráš and Dorčák, 1999). The stability condition can then be stated as follows (Matignon, 1996, 1998; Vinagre and Feliu, 2007):

Theorem 4.1. (Matignon, 1996): *A commensurate-order system described by a rational transfer function* (3.5) *is stable if and only if*

$$|\arg(\lambda_i)| > \alpha \frac{\pi}{2}, \text{ for all } i$$

with λ_i being the i-th root of $P(s^\alpha)$.

For the FOLTI system with commensurate order where the system poles are in general complex conjugate, the stability condition can also be expressed as follows:

Theorem 4.2. (Matignon, 1996, 1998): *A commensurate-order system described by a rational transfer function*

$$G(w) = \frac{Q(w)}{P(w)},$$

where $w = s^q$, $q \in \mathrm{R}^+$, $(0 < q < 2)$, is stable if and only if

$$|\arg(w_i)| > q \frac{\pi}{2},$$

with $\forall w_i \in C$ being the i-th root of $P(w) = 0$.

When $w = 0$ is a single root (singularity in origin) of P, the system cannot be stable. For $q = 1$, this is the classical theorem of pole location in the complex plane: P has no pole in the closed right half plane of the first Riemann sheet. The stability region suggested by this theorem tends to the whole s-plane when q tends to 0, corresponds to the Routh-Hurwitz stability when $q = 1$, and tends to the negative real axis when q tends to 2.

Theorem 4.3. (*Aoun et al.*, 2004; *Matignon*, 1998; *Tavazoeiand Haeri*, 2007*a*, *b*, 2008*a*)*: It has been shown that commensurate system* (3.15) *is stable if the following condition is satisfied* (*also if the triplet* **A**, **B**, **C** *is minimal*)

$$|arg(eig(\mathbf{A}))| > q\frac{\pi}{2}, \tag{4.11}$$

where $0 < q < 2$ *and* $eig(\mathbf{A})$ *represents the eigenvalues of matrix* **A**.

Proposition 4.2. *We can assume that some incommensurate-order systems described by the FODE* (3.13) *or* (3.15) *can be decomposed into the following modal form of the fractional transfer function* (*the so-called Laguerre functions* (*Aoun et al.*, 2007)) :

$$F(s) = \sum_{i=1}^{N}\sum_{k=1}^{n_k} \frac{A_{i,k}}{(s^{q_i} + \lambda_i)^k} \tag{4.12}$$

for some complex numbers $A_{i,k}$, λ_i, *and positive integer* n_k.

A system (4.12) *is BIBO stable if and only if* q_i *and the argument of* λ_i *denoted by* $arg(\lambda_i)$ *in* (4.12) *satisfy the inequalities*

$$0 < q_i < 2 \;\; and \;\; |\arg(\lambda_i)| < \pi\left(1 - \frac{q_i}{2}\right) \;\; for\ all\ i. \tag{4.13}$$

Henceforth, we will restrict the parameters q_i *to the interval* $q_i \in (0,2)$. *For the case* $q_i = 1$ *for all i we obtain a classical stability condition for integer-order system* (*no pole is in right half plane*). *The inequalities* (4.13) *were obtained by applying the stability results given in* (*Akcay and Malti*, 2008; *Matignon*, 1998).

Theorem 4.4. (*Deng et al.*, 2007) : *Consider the following autonomous system for internal stability definition*

$${}_0D_t^{\boldsymbol{q}} x(t) = \mathbf{A}x(t), \;\; x(0) = x_0, \tag{4.14}$$

with $\boldsymbol{q} = [q_1, q_2, \ldots, q_n]^T$ *and its n-dimensional representation:*

$$\begin{aligned}
{}_0D_t^{q_1} x_1(t) &= a_{11}x_1(t) + a_{12}x_2(t) + \cdots + a_{1n}x_n(t),\\
{}_0D_t^{q_2} x_2(t) &= a_{21}x_1(t) + a_{22}x_2(t) + \cdots + a_{2n}x_n(t),\\
&\cdots\cdots\\
{}_0D_t^{q_n} x_n(t) &= a_{n1}x_1(t) + a_{n2}x_2(t) + \cdots + a_{nn}x_n(t),
\end{aligned} \tag{4.15}$$

where all q_i's are rational numbers between 0 and 2. Assume m to be the LCM of the denominators u_i's of q_i's , where $q_i = v_i/u_i$, $v_i, u_i \in Z^+$ for $i = 1,2,\ldots,n$ and we set $\gamma = 1/m$. Define:

$$det \begin{pmatrix} \lambda^{mq_1} - a_{11} & -a_{12} & \ldots & -a_{1n} \\ -a_{21} & \lambda^{mq_2} - a_{22} & \ldots & -a_{2n} \\ \vdots & & & \vdots \\ -a_{n1} & -a_{n2} & \ldots & \lambda^{mq_n} - a_{nn} \end{pmatrix} = 0. \quad (4.16)$$

The characteristic equation (4.16) can be transformed to integer-order polynomial equation if all q_i's are rational number. Then the zero solution of system (4.15) is globally asymptotically stable if all roots λ_i's of the characteristic (polynomial) equation (4.16) satisfy

$$|arg(\lambda_i)| > \gamma\frac{\pi}{2} \text{ for all } i.$$

Denoting λ by s^γ in Eq. (4.16), we get the characteristic equation in the form $det(s^\gamma I - A) = 0$.

Corollary 4.1. *Suppose $q_1 = q_2 = \cdots = q_n \equiv q$, $q \in (0,2)$, all eigenvalues λ of matrix A in (3.17) satisfy $|arg(\lambda)| > q\pi/2$, the characteristic equation becomes $det(s^q I - A) = 0$ and all characteristic roots of the system (3.15) have negative real parts (Deng et al., 2007). This result is Theorem 1 of the paper by Matignon (1996).*

Remark 4.1. : Generally, when we assume $s = |r|e^{j\phi}$, where $|r|$ is the modulus and ϕ is the argument of a complex number in the s-plane, respectively, the transformation $w = s^{\frac{1}{m}}$ to the complex w-plane can be viewed as $s = |r|^{\frac{1}{m}} e^{\frac{j\phi}{m}}$ and thus $|\arg(s)| = m.|\arg(w)|$ and $|s| = |w|^m$. Proof of this statement is obvious.

Stability test procedure for a general fractional-order LTI system can be summarized as follows:

The characteristic equation of a general LTI fractional-order system of the form:

$$a_n s^{\alpha_n} + \cdots + a_1 s^{\alpha_1} + a_0 s^{\alpha_0} \equiv \sum_{i=0}^{n} a_i s^{\alpha_i} = 0 \quad (4.17)$$

may be rewritten as

$$\sum_{i=0}^{n} a_i s^{\frac{u_i}{v_i}} = 0$$

and transformed into the w-plane

$$\sum_{i=0}^{n} a_i w^i = 0, \quad (4.18)$$

with $w = s^{\frac{k}{m}}$, where m is the LCM of v_i. The steps for the stability analysis are (Petráš, 2009; Radwan et al., 2009):

1. For given a_i calculate the roots of Eq. (4.18) and find the absolute phase of all roots $|\phi_w|$.
2. Roots in the primary sheet of the w-plane which have corresponding roots in the s-plane can be obtained by finding all roots which lie in the region $|\phi_w| < \frac{\pi}{m}$ then applying the inverse transformation $s = w^m$ (see Remark 4.1). The region where $|\phi_w| > \frac{\pi}{m}$ is not physical. For testing the roots in desired region the matrix approach (4.1) can be used.
3. The condition for stability is $\frac{\pi}{2m} < |\phi_w| < \frac{\pi}{m}$. Condition for oscillation is $|\phi_w| = \frac{\pi}{2m}$, otherwise the system is unstable (Fig. 4.5(b)). If there is no root in the physical s-plane, the system will always be stable.

Example 4.4. : Let us consider the linear fractional-order LTI system described by the transfer function (Dorčák, 1994; Podlubny, 1999):

$$G(s) = \frac{Y(s)}{U(s)} = \frac{1}{0.8s^{2.2} + 0.5s^{0.9} + 1}, \tag{4.19}$$

where the corresponding FODE has the following form:

$$0.8\,{}_0D_t^{2.2}y(t) + 0.5\,{}_0D_t^{0.9}y(t) + y(t) = u(t) \tag{4.20}$$

with zero initial conditions.

The system (4.20) can be rewritten to its state-space representation $(x_1(t) \equiv y(t))$:

$$\begin{bmatrix} {}_0D^{\frac{9}{10}}x_1(t) \\ {}_0D^{\frac{13}{10}}x_2(t) \end{bmatrix} = \begin{bmatrix} 0 & 1 \\ -1/0.8 & -0.5/0.8 \end{bmatrix} \begin{bmatrix} x_1(t) \\ x_2(t) \end{bmatrix} + \begin{bmatrix} 0 \\ 1/0.8 \end{bmatrix} u(t),$$

$$y(t) = \begin{bmatrix} 1 & 0 \end{bmatrix} \begin{bmatrix} x_1(t) \\ x_2(t) \end{bmatrix}. \tag{4.21}$$

The eigenvalues of the matrix $\mathbf{A}$ are $\lambda_{1,2} = -0.3125 \pm 1.0735j$ and then $|\arg(\lambda_{1,2})| = 1.8541$. Because of various derivative orders in (4.21), Theorem 4.3 cannot be used directly.

The analytical solution of the FODE (4.20) for $u(t) = 0$ obtained from general solution (3.9) has the following form:

$$y(t) = \frac{1}{0.8} \sum_{k=0}^{\infty} \frac{(-1)^k}{k!} \left(\frac{1}{0.8}\right)^k \mathcal{E}_k(t, -\frac{0.5}{0.8}; 2.2 - 0.9, 2.2 + 0.9k). \tag{4.22}$$

In Fig. 4.7 is depicted the analytical solution of the FODE (4.20) where $u(t) = 0$. As we can see in the figure, the solution is stable because $\lim_{t\to\infty} y(t) = 0$.

Let us investigate stability according to the previously described method. The corresponding characteristic equation of the system is:

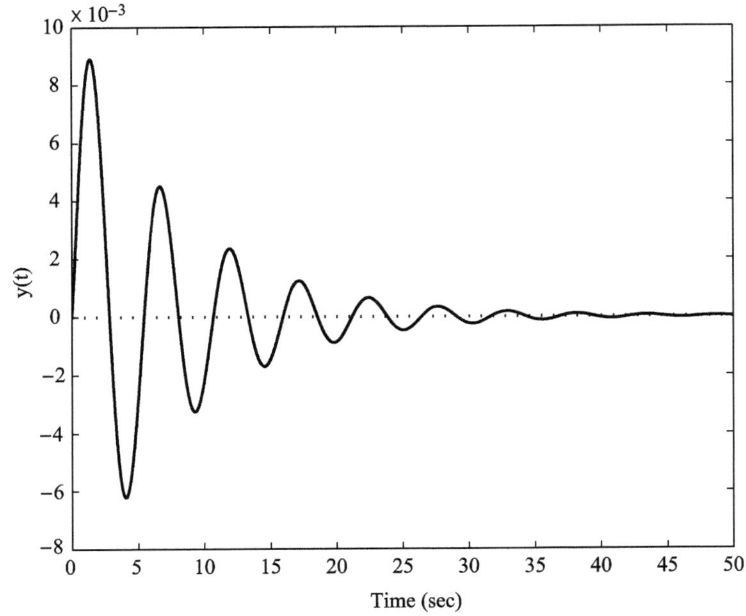

Fig. 4.7 Analytical solution of FODE (4.20) where $u(t) = 0$ for $50s$ and zero initial conditions.

$$P(s) : 0.8s^{2.2} + 0.5s^{0.9} + 1 = 0 \Rightarrow 0.8s^{\frac{22}{10}} + 0.5s^{\frac{9}{10}} + 1 = 0, \qquad (4.23)$$

when $m = 10$, $w = s^{\frac{1}{10}}$, then the roots w_i's and their appropriate arguments of the polynomial

$$P(w) : 0.8w^{22} + 0.5w^9 + 1 = 0 \qquad (4.24)$$

are:

$$
\begin{aligned}
w_{1,2} &= -0.9970 \pm 0.1182j, |\arg(w_{1,2})| = 3.023;\\
w_{3,4} &= -0.9297 \pm 0.4414j, |\arg(w_{3,4})| = 2.698;\\
w_{5,6} &= -0.7465 \pm 0.6420j, |\arg(w_{5,6})| = 2.431;\\
w_{7,8} &= -0.5661 \pm 0.8633j, |\arg(w_{7,8})| = 2.151;\\
w_{9,10} &= -0.259 \pm 0.9625j, |\arg(w_{9,10})| = 1.834;\\
w_{11,12} &= -0.0254 \pm 1.0111j, |\arg(w_{11,12})| = 1.595;\\
w_{13,14} &= 0.3080 \pm 0.9772j, |\arg(w_{11,12})| = 1.265;\\
w_{15,16} &= 0.5243 \pm 0.8359j, |\arg(w_{15,16})| = 1.010;\\
w_{17,18} &= 0.7793 \pm 0.6795j, |\arg(w_{17,18})| = 0.717;\\
w_{19,20} &= 0.9084 \pm 0.3960j, |\arg(w_{19,20})| = 0.411;\\
w_{21,22} &= 1.0045 \pm 0.1684j, |\arg(w_{21,22})| = 0.1661.
\end{aligned}
$$

Physical significance roots are in the first Riemann sheet, which is expressed by relation $-\pi/m < \phi < \pi/m$, where $\phi = \arg(w)$. In this case they are complex conjugate roots $w_{21,22} = 1.0045 \pm 0.1684j$ ($|\arg(w_{21,22})| = 0.1661$), which satisfy conditions $|\arg(w_{21,22})| > \pi/2m = \pi/20$. It means that system (4.20) is stable (Fig. 4.8). Other roots of the polynomial equation (4.24) lie in region $|\phi| > \frac{\pi}{m}$ which is not physical (outside of closed angular sector limited by thick line in Fig. 4.8(b)).

In Fig. 4.8(a) is depicted the Riemann surface of the function $w = s^{\frac{1}{10}}$ with the 10-Riemann sheets and in Fig. 4.8(b) are depicted the roots in complex w-plane with angular sector corresponding to stability region (dash line) and the first Riemann sheet (thick line).

The interesting notion of Remark 4.1 should be mentioned here. The characteristic equation (4.23) has the following poles: $s_{1,2} = -0.10841 \pm 1.19699j$, in the first Riemann sheet in s-plane, which can be obtained, e.g., via the Matlab routine as, for instance, `s=solve('0.8*s^2.2+0.5*s^0.9+1=0','s')`. When we compare $|\arg(w_{21,22})| = 0.1661$ with $|\arg(s_{1,2})| = 1.661$, we can see that $|\arg(s_{1,2})| = m|\arg(w_{21,22})|$, where $m = 10$ in transformation $w = s^{\frac{1}{m}}$. The first Riemann sheet is transformed from the s-plane to the w-plane as follows: $-\pi/10 < \arg(w) < \pi/10$ and then $-\pi < 10\arg(w) < \pi$. Therefore from this consideration we obtain relation $|\arg(s)| = 10\,|\arg(w)|$.

Example 4.5. : Consider the closed loop system with controlled system (electrical heater)

$$G(s) = \frac{1}{39.96s^{1.25} + 0.598} \tag{4.25}$$

and PD controller

$$C(s) = 64.47 + 12.46s. \tag{4.26}$$

The resulting closed loop transfer function $G_c(s)$ becomes (Petráš et al., 2004b):

$$G_c(s) = \frac{Y(s)}{W(s)} = \frac{12.46s + 64.47}{39.69s^{1.25} + 12.46s + 65.068} \tag{4.27}$$

The analytical solution (impulse response) of the fractional-order control system (4.27) is:

$$y(t) = \frac{12.46}{39.69}\sum_{k=0}^{\infty}\frac{(-1)^k}{k!}\left(\frac{12.46}{39.69}\right)^k \times \mathscr{E}_k(t, -\frac{65.068}{39.69}; 1.25, 0.25-k) \tag{4.28}$$
$$+\frac{64.47}{39.69}\sum_{k=0}^{\infty}\frac{(-1)^k}{k!}\left(\frac{65.068}{39.69}\right)^k \times \mathscr{E}_k(t, -\frac{12.46}{39.69}; 1.25-1, 1.25+k)$$

with zero initial conditions.

The characteristic equation of this system is

$$39.69s^{1.25} + 12.46s + 65.068 = 0 \Rightarrow 39.69s^{\frac{5}{4}} + 12.46s^{\frac{4}{4}} + 65.068 = 0 \tag{4.29}$$

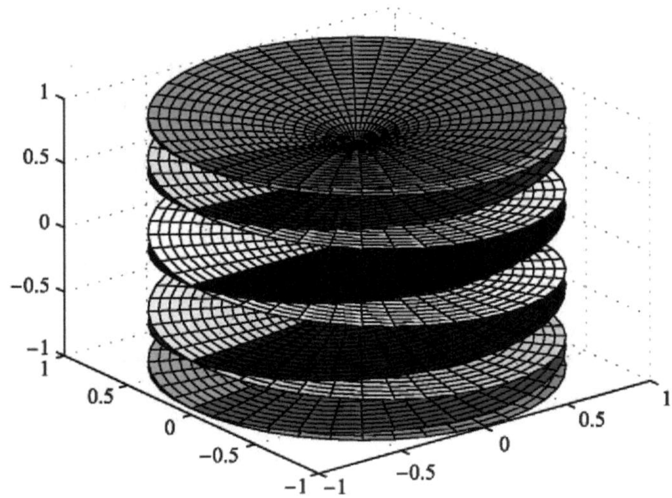

(a) 10-sheets Riemann surface

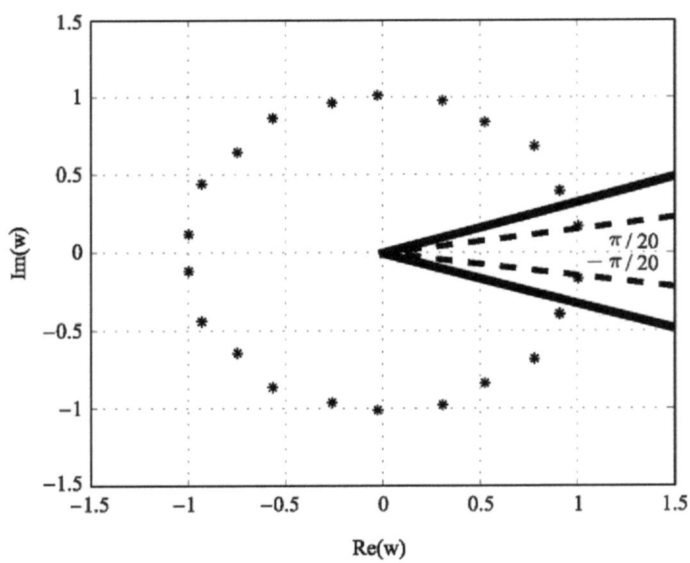

(b) Poles in complex w-plane

Fig. 4.8 Riemann surface of function $w = s^{\frac{1}{10}}$ and roots of Eq. (4.24) in complex w-plane.

Using the notation $w = s^{\frac{1}{m}}$, where LCM is $m = 4$, we obtain a polynomial of complex variable w in the form

$$39.69w^5 + 12.46w^4 + 65.068 = 0. \tag{4.30}$$

Solving the polynomial (4.30) we get the following roots and their arguments:

$$w_1 = -1.17474, |\arg(w_1)| = \pi,$$

$$w_{2,3} = -0.40540 \pm 1.0426j, |\arg(w_{2,3})| = 1.9416,$$
$$w_{4,5} = 0.83580 \pm 0.64536j, |\arg(w_{4,5})| = 0.6575.$$

This first Riemann sheet is defined as a sector in w-plane within interval $-\pi/4 < \arg(w) < \pi/4$. Complex conjugate roots $w_{4,5}$ lie in this interval and satisfy the stability condition given as $|\arg(w)| > \frac{\pi}{8}$, therefore the system is stable. The region where $|\arg(w)| > \frac{\pi}{4}$ is not physical.

Example 4.6. : Let us examine an interesting example of application, the so-called Bessel function of the first kind, the transfer function of which is (Matignon, 1998):

$$H(s) = \frac{1}{\sqrt{s^2+1}}, \quad \forall s, \ \Re(s) > 0. \tag{4.31}$$

We have two branch points $s_1 = j$, and $s_2 = -j$ and two cuts. One along the half line $(-\infty + j, j)$ and the other along the half line $(-\infty - j, -j)$. In this doubly cut complex plane, we have the identity $\sqrt{s^2+1} = \sqrt{s-j}\sqrt{s+j}$. The well-known asymptotic expansion of Eq. (4.31) is:

$$h(t) \approx \sqrt{\frac{2}{\pi t}} \cos(t - \frac{\pi}{4}) = \sqrt{\frac{2}{\pi}} t^{-\frac{1}{2}} E_{2,1}\left(-(t - \frac{\pi}{4})^2\right).$$

According to the branch points and above asymptotic expansion we can state that the system described by the Bessel function (4.31) is on the boundary of stability and has oscillation behaviour.

Example 4.7. : Let us consider the more complex example, namely, Nuclear Magnetic Resonance (NMR). In physics and bio-engineering, specifically in NMR or magnetic resonance imaging the Bloch equations are a set of macroscopic equations that are used to calculate the nuclear magnetization $\mathbf{M} = (M_x(t), M_y(t), M_z(t))$ as a function of time when relaxation time is T_1 (spin-lattice) and T_2 (spin-spin). These equations were introduced by Felix Bloch in 1946 and can be expressed in the following form (Bloch, 1946):

$$\begin{aligned}
\frac{dM_x(t)}{dt} &= \gamma(\mathbf{M}(t) \times \mathbf{B}(t))_x - \frac{M_x(t)}{T_2}, \\
\frac{dM_y(t)}{dt} &= \gamma(\mathbf{M}(t) \times \mathbf{B}(t))_y - \frac{M_y(t)}{T_2},
\end{aligned} \tag{4.32}$$

$$\frac{dM_z(t)}{dt} = \gamma(\mathbf{M}(t) \times \mathbf{B}(t))_z - \frac{M_z(t) - M_0}{T_1},$$

where $\gamma/2\pi$ is the gyromagnetic ratio, $\mathbf{B}(t) = (B_x(t), B_y(t), B_0 + \Delta B_z(t))$ is the magnetic field experienced by the nuclei, and M_0 is the equilibrium magnetization.

However, the relaxation terms describe the return to equilibrium, but only for a field pointing along the z-axis, the Bloch equations (4.32) for the constant static magnetic field B_0 (z-component) reduce to the equations (Haacke et al., 1999):

$$\begin{aligned} \frac{dM_x(t)}{dt} &= \omega_0 M_y(t) - \frac{M_x(t)}{T_2}, \\ \frac{dM_y(t)}{dt} &= -\omega_0 M_x(t) - \frac{M_y(t)}{T_2}, \\ \frac{dM_z(t)}{dt} &= \frac{M_0 - M_z(t)}{T_1}, \end{aligned} \tag{4.33}$$

where $\omega_0 = \gamma B_0$ and $\omega_0 = 2\pi f_0$ (e.g. gyromagnetic ratio $\gamma/2\pi = f_0/B_0 = 42.57$ Mhz/Tesla for water protons).

Now, we consider the fractional-order Bloch equations, where integer-order derivatives are replaced by fractional-order ones. The mathematical description of the fractional-order system with Caputo's derivatives is expressed as (Magin et al., 2009):

$$\begin{aligned} {}_0D_t^{q_1} M_x(t) &= \omega_0' M_y(t) - \frac{M_x(t)}{T_2'}, \\ {}_0D_t^{q_2} M_y(t) &= -\omega_0' M_x(t) - \frac{M_y(t)}{T_2'}, \\ {}_0D_t^{q_3} M_z(t) &= \frac{M_0 - M_z(t)}{T_1'}, \end{aligned} \tag{4.34}$$

where q_1, q_2, and q_3 are the derivative orders. The total order of the system is $\bar{q}$=(q_1, q_2, q_3). Here, ω_0', T_1', and T_2' have the units of $(\mathrm{sec})^{-q}$ to maintain a consistent set of units for the magnetization.

Analytical solution of the fractional-order Bloch equations (4.34) based on Mittag-Leffler function has been derived and discussed in (Magin et al., 2009). In addition, we will use a more convenient numerical solution of the fractional-order Bloch equations (4.34), which is based on the Grünwald-Letnikov method described in Chapter 2, having the form:

$$M_x(t_k) = \left(\omega_0' M_y(t_{k-1}) - \frac{M_x(t_{k-1})}{T_2'} \right) h^{q_1} - \sum_{j=v}^{k} c_j^{(q_1)} M_x(t_{k-j}),$$

$$M_y(t_k) = \left(-\omega_0^{'} M_x(t_k) - \frac{M_y(t_{k-1})}{T_2^{'}}\right) h^{q_2} - \sum_{j=v}^{k} c_j^{(q_2)} M_y(t_{k-j}), \qquad (4.35)$$

$$M_z(t_k) = \left(\frac{M_0 - M_z(t_{k-1})}{T_1^{'}}\right) h^{q_3} - \sum_{j=v}^{k} c_j^{(q_3)} M_z(t_{k-j}),$$

where T_{sim} is the simulation time, $k = 1,2,3,\ldots,N$, for $N = [T_{sim}/h]$, and $(M_x(0), M_y(0), M_z(0))$ is the start point (initial conditions). The binomial coefficients $c_j^{(q_i)}\ \forall i$ are calculated according to relation (2.54). All simulations described in this section were performed without using the short memory principle ($v = 1$) for time step $h = 0.00001$.

Stability of the fractional-order Bloch equations (4.34) can be investigated according to Theorem 4.3 or Theorem 4.4. The first and second equations of set (4.34) are coupled and the third one is independent of them. The stability condition is determined from the following expression

$$_0D_t^{\mathbf{q}} \begin{bmatrix} M_x(t) \\ M_y(t) \end{bmatrix} = \begin{bmatrix} -\frac{1}{T_2^{'}} & \omega_0^{'} \\ -\omega_0^{'} & -\frac{1}{T_2^{'}} \end{bmatrix} \begin{bmatrix} M_x(t) \\ M_y(t) \end{bmatrix}, \qquad (4.36)$$

where $\mathbf{q} = [q_1, q_2]^T$.

The system matrix is defined as

$$\mathbf{A} = \begin{bmatrix} -\dfrac{1}{T_2{'}} & \omega_0^{'} \\ -\omega_0^{'} & -\dfrac{1}{T_2{'}} \end{bmatrix}. \qquad (4.37)$$

For the following system parameters (Magin et al., 2009): $T_2^{'} = 20\,(\mathrm{ms})^q$, and $f_0 = 160\,\mathrm{Hz}$ we obtain the eigenvalues $\mathrm{eig}(\mathbf{A}) = -50 \pm 1005.3j$ and $|\arg(\mathrm{eig}(\mathbf{A}))| = 1.6205$. According to the stability condition of Theorem 4.3, the system (4.36) for the above parameters is *stable* if $q < 1.03163$ in the case $q_1 = q_2$. For $q_1 = q_2 \approx 1.03163$ we get the critical stability border and the solution of the system (4.36) is depicted in Fig. 4.9. In Fig. 4.9(a), we observe a limit cycle and Fig. 4.9(b) plots spiral.

In the case where we consider $q_1 = q_2 = q_3 = 1$ in (4.34), we have the integer-order (classical) model of the Bloch equations (4.32), and the numerical solution obtained by (4.35) is shown in Fig. 4.10.

When we consider $q_1 = q_2 = q_3 = 0.9$ in (4.34), we have the fractional-order model of the Bloch equations (4.32), and the numerical solution obtained by (4.35) is shown in Fig. 4.11.

When we consider $q_1 = 0.8$, $q_2 = 0.9$, and $q_3 = 1.0$ in (4.34), we have the fractional-order model of the Bloch equations (4.32), and the numerical solution obtained by (4.35) is shown in Fig. 4.12.

According to Theorem 4.4, the stability condition for equations orders $q_1 = 0.8$, $q_2 = 0.9$, $q_3 = 1.0$ of the solution depicted in Fig. 4.12 is given as $|\arg(\lambda_i)| >$

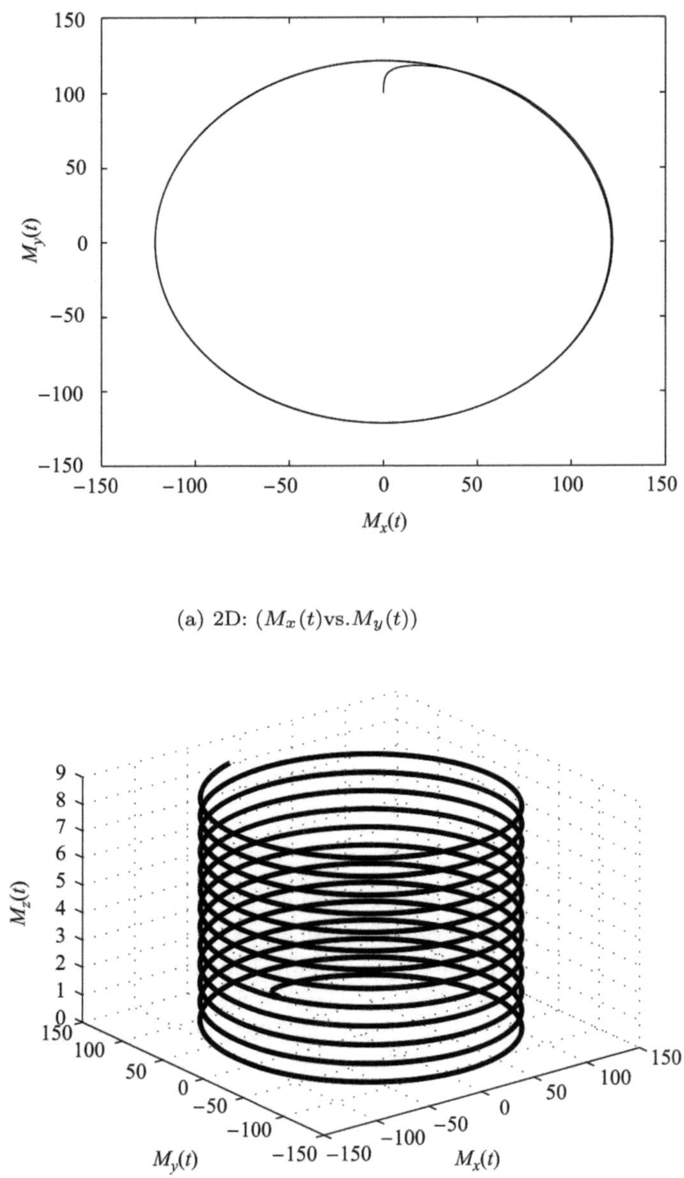

(a) 2D: ($M_x(t)$vs.$M_y(t)$)

(b) 3D: ($M_x(t)$vs.$M_y(t)$vs.$M_z(t)$)

Fig. 4.9 Numerical solutions of Bloch equations (4.35) with parameters: $q \equiv q_1 = q_2 \approx 1.03163$, $T_1^{'} = 1\,(s)^q$, $T_2^{'} = 20\,(\mathrm{ms})^q$, $f_0 = 160\,\mathrm{Hz}$, and initial conditions $M_x(0) = 0$, $M_y(0) = 100$, $M_z(0) = 0$ for $T_{sim} = 0.1\,s$.

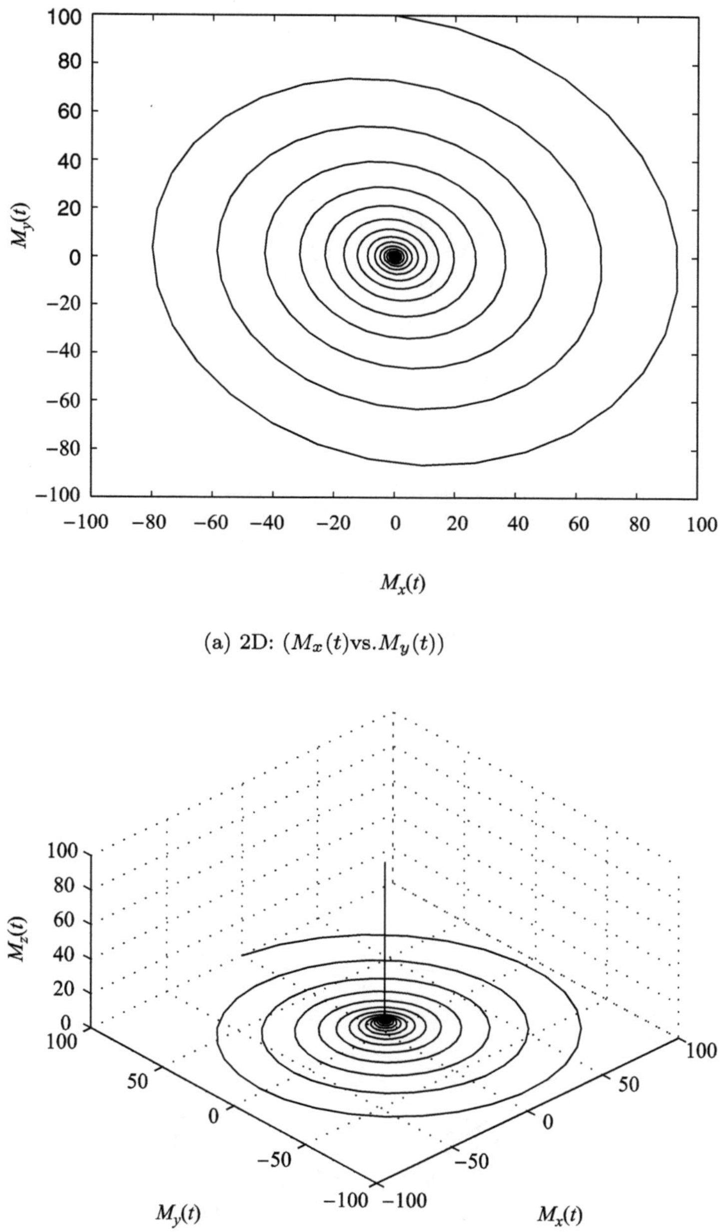

(a) 2D: ($M_x(t)$vs.$M_y(t)$)

(b) 3D: ($M_x(t)$vs.$M_y(t)$vs.$M_z(t)$)

Fig. 4.10 Numerical solutions of Bloch equations (4.34) with parameters: $q \equiv q_1 = q_2 = q_3 = 1$, $T_1^{'} = 1\,s$, $T_2^{'} = 20\,\text{ms}$, $f_0 = 160\,\text{Hz}$, and initial conditions $M_x(0) = 0$, $M_y(0) = 100$, $M_z(0) = 0$ for $T_{sim} = 1\,s$

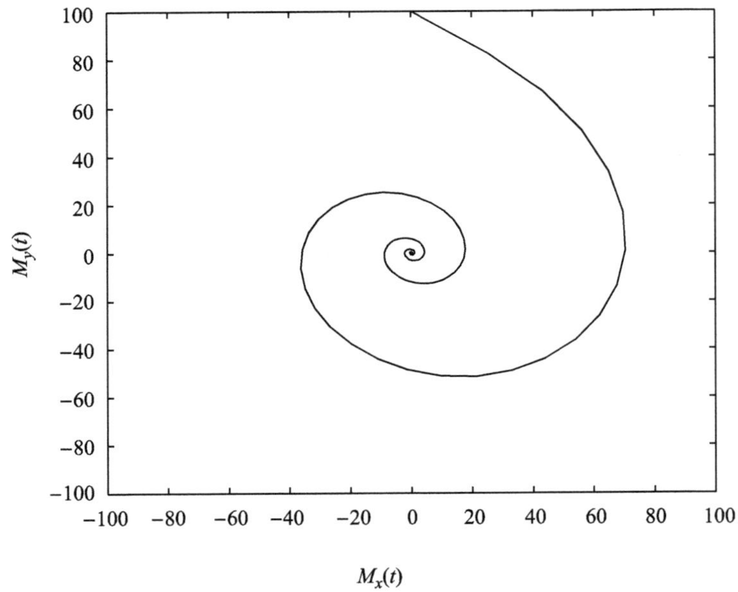

(a) 2D: ($M_x(t)$vs.$M_y(t)$)

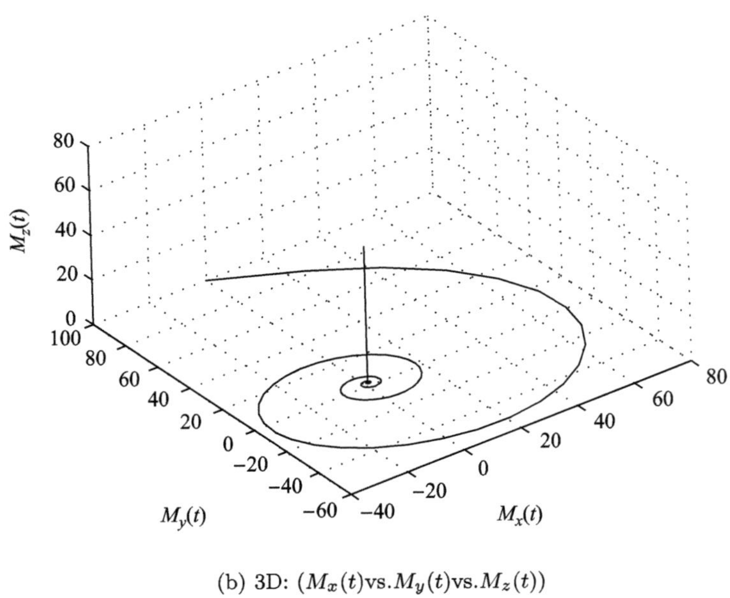

(b) 3D: ($M_x(t)$vs.$M_y(t)$vs.$M_z(t)$)

Fig. 4.11 Numerical solutions of Bloch equations (4.35) with parameters: $q \equiv q_1 = q_2 = q_3 = 0.9$, $T_1^{'} = 1\,s$, $T_2^{'} = 20\,\mathrm{ms}$, $f_0 = 160\,\mathrm{Hz}$, and initial conditions $M_x(0) = 0$, $M_y(0) = 100$, $M_z(0) = 0$ for $T_{sim} = 1\,s$.

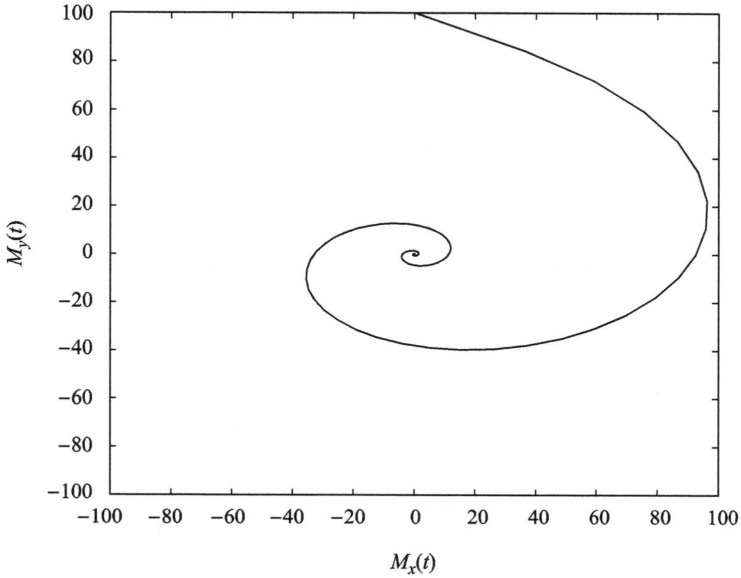

(a) 2D: $(M_x(t)\text{vs.}M_y(t))$

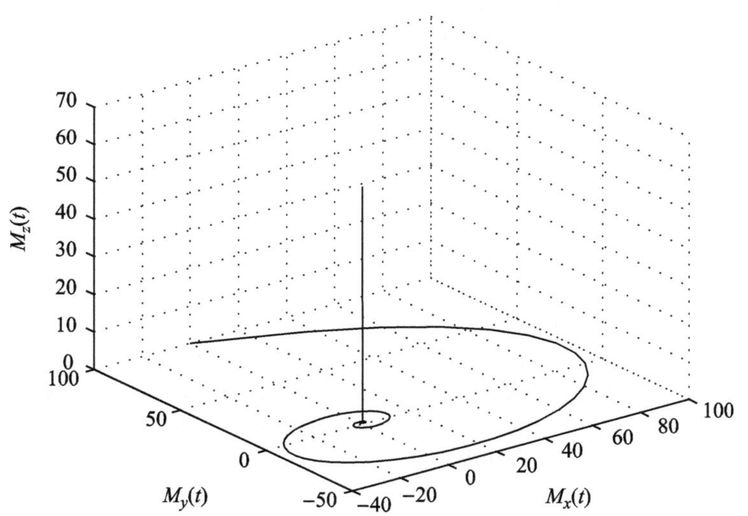

(b) 3D: $(M_x(t)\text{vs.}M_y(t)\text{vs.}M_z(t))$

Fig. 4.12 Numerical solutions of Bloch equations (4.35) with parameters: $q_1 = 0.8$, $q_2 = 0.9$, $q_3 = 1.0$, $T_1' = 1\,s$, $T_2' = 20\,\text{ms}$, $f_0 = 160\,\text{Hz}$, and initial conditions $M_x(0) = 0$, $M_y(0) = 100$, $M_z(0) = 0$ for $T_{sim} = 1\,s$.

$\pi/(2m)$, $\forall i$. For $m = 10$ we get the characteristic polynomial in the form

$$\lambda^{17} + 50\lambda^{8} + 50\lambda^{9} + 2500 + 102400\pi^2 = 0.$$

All λ_i $(i = 1,2,\ldots,17)$ satisfy the condition $|\arg(\lambda_i)| > \pi/20$ and therefore the system is *stable.*

Example 4.8. : Let us consider two examples of certain class of the fractional-order systems (Kheirizad et al., 2009). The first of them has the following form:

$$
{}_0D_t^{1.1}\begin{bmatrix} x_1(t) \\ x_2(t) \\ x_3(t) \end{bmatrix} = \begin{bmatrix} -8 & 2 & -3 \\ -1 & -2 & 0.2 \\ 0.5 & -1 & -2 \end{bmatrix}\begin{bmatrix} x_1(t) \\ x_2(t) \\ x_3(t) \end{bmatrix} + \begin{bmatrix} 1 \\ 1 \\ 1 \end{bmatrix} u(t), \tag{4.38}
$$

$$
y(t) = \begin{bmatrix} 0 \; 0 \; 1 \end{bmatrix}\begin{bmatrix} x_1(t) \\ x_2(t) \\ x_3(t) \end{bmatrix},
$$

where $x \in \mathrm{R}^3$. For the system matrix $\mathbf{A}$ of the system (4.38) we get the eigenvalues $\lambda_{1,2} = -2.2719 \pm 0.8119j, \lambda_3 = -7.4562$. All eigenvalues satisfy the stability conditions of Theorem 4.3, where $|\arg(\lambda_{1,2})| = 2.7984$ and $|\arg(\lambda_3)| = \pi$, thus $|\arg(eig(\mathbf{A}))| > 1.1\frac{\pi}{2}$ and therefore the system (4.38) is *stable.*

The second of them has the following form:

$$
\begin{aligned}
{}_0D_t^{1.2}\begin{bmatrix} x_1(t) \\ x_2(t) \\ x_3(t) \end{bmatrix} &= \begin{bmatrix} -4 & 1 & 1 \\ 0 & -3 & 1 \\ 1 & 3 & 1 \end{bmatrix}\begin{bmatrix} x_1(t) \\ x_2(t) \\ x_3(t) \end{bmatrix} + \begin{bmatrix} 0 \\ 0 \\ 1 \end{bmatrix} u(t), \\
y(t) &= \begin{bmatrix} 0 \; 0 \; 1 \end{bmatrix}\begin{bmatrix} x_1(t) \\ x_2(t) \\ x_3(t) \end{bmatrix}.
\end{aligned} \tag{4.39}
$$

For the system matrix $\mathbf{A}$ of the system (4.39) we get the eigenvalues $\lambda_1 = 1.8284, \lambda_2 = -4$, and $\lambda_3 = -3.8284$. The eigenvalue λ_1 does not satisfy the stability condition of Theorem 4.3, where $|\arg(\lambda_1)| = 0$ and $|\arg(\lambda_{2,3})| = \pi$, thus $|\arg(\lambda_1)| < 1.2\frac{\pi}{2}$ and therefore system (4.39) is *unstable.*

4.3 Stability of Fractional Nonlinear Systems

Stability of the fractional-order nonlinear system is very complex and is different from the fractional-order linear system. The main difference is that for a nonlinear system it is necessary to investigate steady states having two types: equilibrium point and limit cycle. Nonlinear systems may have several equlibrium points. For nonlinear systems, there are many definitions of stability (asymptotic, global, orbital, etc.). The basic idea was formulated by A. M. Lyapunov.

As mentioned in (Matignon, 1996), exponential stability cannot be used to characterize asymptotic stability of fractional-order systems. A new definition was introduced (Oustaloup et al., 2008).

Definition 4.4. Trajectory $x(t) = 0$ of the system (3.19) is t^{-q} asymptotically stable if there is a positive real q so that:

$$\forall ||x(t)|| \text{ with } t \leq t_0, \exists N(x(t)), \text{ such that } \forall t \geq t_0, ||x(t)|| \leq N t^{-q}.$$

The fact that the components of $x(t)$ slowly decay towards 0 following t^{-q} leads to fractional systems sometimes called long memory systems. Power law stability t^{-q} is a special case of the Mittag-Leffler stability (Li et al., 2008).

Theorem 4.5. *According to stability theorem defined in* (*Tavazoei and Haeri,* 2008b), *the equilibrium points are asymptotically stable for* $q_1 = q_2 = \cdots = q_n \equiv q$ *if all the eigenvalues* λ_i, $(i = 1, 2, \ldots, n)$ *of the Jacobian matrix* $\boldsymbol{J} = \partial \boldsymbol{f}/\partial \boldsymbol{x}$, *where* $\boldsymbol{f} = [f_1, f_2, \ldots, f_n]^T$, *evaluated at the equilibrium* E^*, *satisfy the condition* (*Tavazoei and Haeri,* 2007b,a):

$$|\arg(\text{eig}(\boldsymbol{J}))| = |\arg(\lambda_i)| > q\frac{\pi}{2}, \; i = 1, 2, \ldots, n. \tag{4.40}$$

Figure 4.6 *shows stable and unstable regions of the complex plane for such case.*

Theorem 4.6. *When we consider the incommensurate fractional-order system* $q_1 \neq q_2 \neq \cdots \neq q_n$ *and suppose that m is the LCM of the denominators* u_i*'s of* q_i*'s, where* $q_i = v_i/u_i$, $v_i, u_i \in Z^+$ *for* $i = 1, 2, \ldots, n$ *and we set* $\gamma = 1/m$. *System* (3.20) *is asymptotically stable if*

$$|\arg(\lambda)| > \gamma\frac{\pi}{2}$$

for all roots λ *of the following equation*

$$\det(\text{diag}([\lambda^{mq_1} \, \lambda^{mq_2} \cdots \lambda^{mq_n}]) - \boldsymbol{J}) = 0. \tag{4.41}$$

Proof. The proof of this statement is obvious (Tavazoei and Haeri, 2008b).

A necessary stability condition for fractional-order systems (3.20) to remain chaotic is keeping at least one eigenvalue λ in the unstable region (Tavazoei and Haeri, 2007b). The number of equilibrium points and eigenvalues for one-scroll, double-scroll and multi-scroll attractors was exactly described in the work by Tavazoei and Haeri (2008a). Assume that a 3D chaotic system has only three equilibria. Therefore, if the system has a double-scroll attractor, it has two saddle-focus points surrounded by scrolls and one additional saddle point.

Definition 4.5. Suppose that the unstable eigenvalues of scroll focus points are: $\lambda_{1,2} = \alpha_{1,2} \pm j\beta_{1,2}$. The necessary condition to exhibit double-scroll attractor of system (3.20) is the eigenvalues $\lambda_{1,2}$ remaining in the unstable region (Tavazoei and Haeri, 2008a). The condition for commensurate derivatives order is

$$q > \frac{2}{\pi}\operatorname{atan}\left(\frac{|\beta_i|}{\alpha_i}\right), \quad i = 1,2. \tag{4.42}$$

This condition can be used to determine the minimum order for which a nonlinear system can generate chaos (Tavazoei and Haeri, 2007b). In other words, when the instability measure $\pi/2m - \min(|\arg(\lambda)|)$ is negative, the system cannot be chaotic.

Example 4.9. : Let us investigate the singular points of the nonlinear systems (Kotek et al., 1973):

$$\begin{aligned}\frac{dx_1(t)}{dt} &= x_2(t),\\ \frac{dx_2(t)}{dt} &= bx_1(t) - cx_1^3(t) - ax_2(t),\end{aligned} \tag{4.43}$$

for $b, c > 0$. The system (4.43) has three singular points (equilibrium) at $E_1 = (0,0)$, $E_2 = (+\sqrt{\frac{b}{c}},0)$, and $E_3 = (-\sqrt{\frac{b}{c}},0)$. The Jacobian matrix of the system evaluated at equilibrium $E^* = (x_1^*, x_2^*)$ is:

$$\mathbf{J} = \begin{bmatrix} 0 & 1 \\ b - 3cx_1^{*2} & -a \end{bmatrix}. \tag{4.44}$$

For the first singular point ($x_1^* = 0, x_2^* = 0$) we have the following characteristic equation

$$\det(\lambda\mathbf{I} - \mathbf{J}) = \lambda^2 + a\lambda - b = 0,$$

where the real roots are: $\lambda_{1,2} = (-a \pm \sqrt{a^2 + 4b})/2$. The singular point is a saddle point.

For the second and third singular points $E_2 = (+\sqrt{\frac{b}{c}},0)$, and $E_3 = (-\sqrt{\frac{b}{c}},0)$ we have the following characteristic equation

$$\det(\lambda\mathbf{I} - \mathbf{J}) = \lambda^2 + a\lambda + 2b = 0,$$

where the roots are: $\lambda_{1,2} = (-a \pm \sqrt{a^2 - 8b})/2$. The singular point depends on the values of a and b. For $(a^2/4) < 2b$ we get complex-conjugate roots and the singular point is a stable focus. For $(a^2/4) > 2b$ we get real negative roots and the singular point is a stable node. It is valid for 2D systems. For 3D systems it is a bit different.

When the type of singular point is known, it is possible to construct approximated shape of a phase trajectory around the singular points. From the theory of stability for nonlinear systems, for the above-mentioned singular points, the net of phase trajectories in the (x_1, x_2) plane resembles the trajectories for a flip-flop circuit with two stable steady states.

Example 4.10. : Let us investigate Chen's system with a double scroll attractor in 3D state space. The fractional-order form of such system can be described as (Tavazoei and Haeri, 2008b):

$$
\begin{aligned}
{}_0D_t^{0.8}x_1(t) &= 35[x_2(t) - x_1(t)], \\
{}_0D_t^{1.0}x_2(t) &= -7x_1(t) - x_1(t)x_3(t) + 28x_2(t), \\
{}_0D_t^{0.9}x_3(t) &= x_1(t)x_2(t) - 3x_3(t).
\end{aligned}
\tag{4.45}
$$

The system has three equilibria at $(0,0,0)$, $(7.94, 7.94, 21)$, and $(-7.94, -7.94, 21)$. The Jacobian matrix of the system evaluated at equilibrium $E^* = (x_1^*, x_2^*, x_3^*)$ is:

$$
\mathbf{J} = \begin{bmatrix} -35 & 35 & 0 \\ -7 - x_3^* & 28 & -x_1^* \\ x_2^* & x_1^* & -3 \end{bmatrix}. \tag{4.46}
$$

The last two equilibrium points are saddle points and are surrounded by a chaotic double scroll attractor. For these two points, Equation (4.41) becomes:

$$
\lambda^{27} + 35\lambda^{19} + 3\lambda^{18} - 28\lambda^{17} + 105\lambda^{10} - 21\lambda^{8} + 4410 = 0. \tag{4.47}
$$

The characteristic equation (4.47) has unstable roots $\lambda_{1,2} = 1.2928 \pm 0.2032j$, $|\arg(\lambda_{1,2})| = 0.1560$, therefore the system (4.45) satisfies the necessary condition for exhibiting a double scroll attractor. Instability measure is 0.0012.

Numerical simulation of the system (4.45) for initial conditions $(-9, -5, 14)$ is depicted in Fig. 4.13.

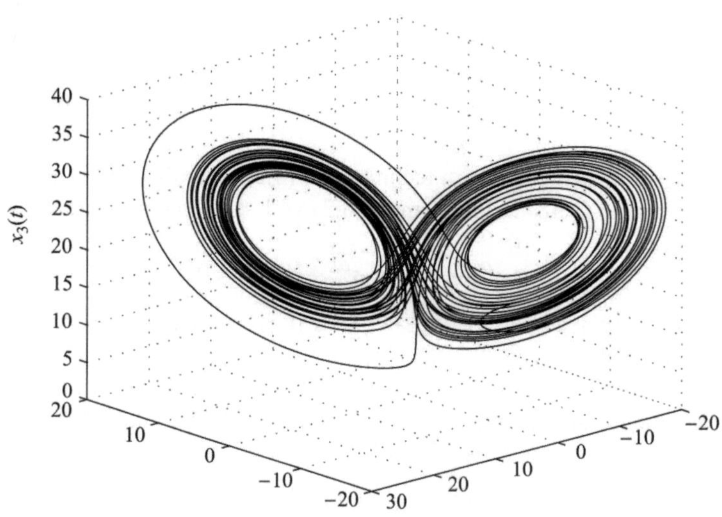

Fig. 4.13 Double scroll attractor of Chen's system (4.45) projected into 3D state space for 30 *s*.

4.4 Robust Stability of Fractional-Order LTI Systems

In this section, we consider an even more complicated situation where the parameters, even the fractional orders in (3.2) are not exactly known but are interval real numbers with known interval bounds.

The motivation of this problem comes from the modelling of the dynamics of nonferrous materials and/or their composites such as the nonlaminated magnetic bearing. The fractional order of the LTI model may be derived from the mean slope of the Bode plot. The variance of the slopes of the measured Bode plots can be practically described by using an interval of width $\pm 3\gamma$ centered at the mean slope.

For the LTI FOS with uncertainty several methods were developed for its stability analysis (Chen et al., 2006b; Petráš et al., 2004a,b; Tan et al., 2009). A method based on existence of a Hermitian matrix that the complex Lyapunov inequalities are satisfied for all vertex matrices was described in (Ahn and Chen, 2008). A linear matrix inequality (LMI) approach was given in (Lu and Chen, 2009).

4.4.1 Stability Check When the Fractional Orders Are Crisp and Commensurate

Consider the following LTI FOS with interval uncertainties in the parameters:

$$ {}_0D_t^{\alpha} x(t) = Ax(t) + Bu(t), \tag{4.48}$$

where α is the fractional commensurate order, $x(t) \in \mathrm{R}^n$ and $u(t) \in \mathrm{R}^p$ denote the state and the control vector, respectively, the system matrices A and B are interval uncertain in parameterwise, defined as $A \in A_I = [\underline{A}, \bar{A}]$, and $B \in B_I = [\underline{B}, \bar{B}]$, where $\underline{A} = [\underline{a_{ij}}]_{n\times n}$, $\bar{A} = [\bar{a_{ij}}]_{n\times n}$ satisfy $\underline{a_{ij}} \le \bar{a_{ij}}$, and $\underline{B} = [\underline{b_{ij}}]_{n\times p}$, $\bar{B} = [\bar{b_{ij}}]_{n\times p}$ satisfy $\underline{b_{ij}} \le \bar{b_{ij}}$ for all $1 \le i \le n,\ 1 \le j \le p$.

Applying the Laplace trasform technique to (4.48) we obtain a transfer function $G(s)$ of the system (4.48) in the form of (3.2) with the interval characteristic polynomial:

$$a_n s^{\alpha_n} + \cdots + a_1 s^{\alpha_1} + a_0 s^{\alpha_0} = 0,$$

where s is the Laplace operator, $\alpha_i > \alpha_j > 0$ if $i < j$, and $a_i \in [\underline{a_i}, \bar{a}_i]$ are interval parameters beyond $[q_i^-, q_i^+]$.

The *robust stability* test procedure for the LTI FOS of commensurate orders with interval parametric uncertainties was proposed in (Petráš et al., 2002a) and can be divided into the following steps:

- **step 1:** Transcribe the LTI FOS $G(s)$ of the commensurate order α into the equivalent system $H(\sigma)$, where the transformation is: $s^{\alpha} \to \sigma$, $\alpha \in \mathrm{R}^+$;
- **step 2:** Write the interval polynomial $P(\sigma, q)$ of the equivalence system $H(\sigma)$, where interval polynomial is defined as

$$P(\sigma,q)=\sum_{i=0}^{n}[q_i^-,q_i^+]\sigma^i;$$

- **step 3:** For interval polynomial $P(\sigma,q)$, construct four Kharitonov's polynomials: $p^{--}(\sigma),p^{-+}(\sigma),p^{+-}(\sigma),p^{++}(\sigma)$;
- **step 4:** Test the four Kharitonov's polynomials to clarify whether they satisfy the stability condition: $|\arg(\sigma_i)|>\alpha\frac{\pi}{2}$, $\forall\sigma\in C$, with σ_i being the i-th root of $P(\sigma)$;

Note that for low-degree polynomials, fewer Kharitonov's polynomials are to be tested (Henrion, 2001; Yeung and Wang, 1987):

- Degree 5: $p^{--}(\sigma),p^{-+}(\sigma),p^{+-}(\sigma)$;
- Degree 4: $p^{+-}(\sigma),p^{++}(\sigma)$;
- Degree 3: $p^{+-}(\sigma)$.

We demonstrate this technique for the robust stability check for the LTI FOS with parametric interval uncertainties through the following illustrative example (see also (Petráš et al., 2002a) for the time-domain and the frequency cross-validation).

Example 4.11. : Consider a family of the LTI FOS (with commensurate orders) described by

$$G(s,a,b)=\frac{1}{s^{1.5}+as^{0.5}+b},\tag{4.49}$$

where $a\in[0,1]$ and $b\in[1,2]$.
Question: Is $G(s,a,b)$ robustly stable for all a and b ?
The unit-step response is found by (3.9):

$$y(t)=\sum_{k=0}^{\infty}\frac{(-1)^k}{k!}(b)^k\,\mathscr{E}_k(t,-a;1,1.5+0.5k+1)\tag{4.50}$$

with zero initial conditions.

Graphical Method:

For the interval uncertainty of the parameters a and b we can show the area of roots for characteristic equation of the system (4.49) in the s-complex plane. As we can see in Fig. 4.14, all roots are located in the left half plane.

Fig. 4.15 shows the unit step responses to various values of parameter $a\in[0,1]$ with a fixed value of variable b $(b=1)$ and to time $T_{sim}=7\,s$.

Fig. 4.16 shows the unit step responses to various values of parameter $a\in[0,1]$ with a fixed value of variable b $(b=2)$ and to time $T_{sim}=7\,s$.

For better visibility of robust stability, Fig. 4.17 is presented to show the 3D projection of the stable plane for characteristic pseudo-polynomial with various values of the coefficients a and b.

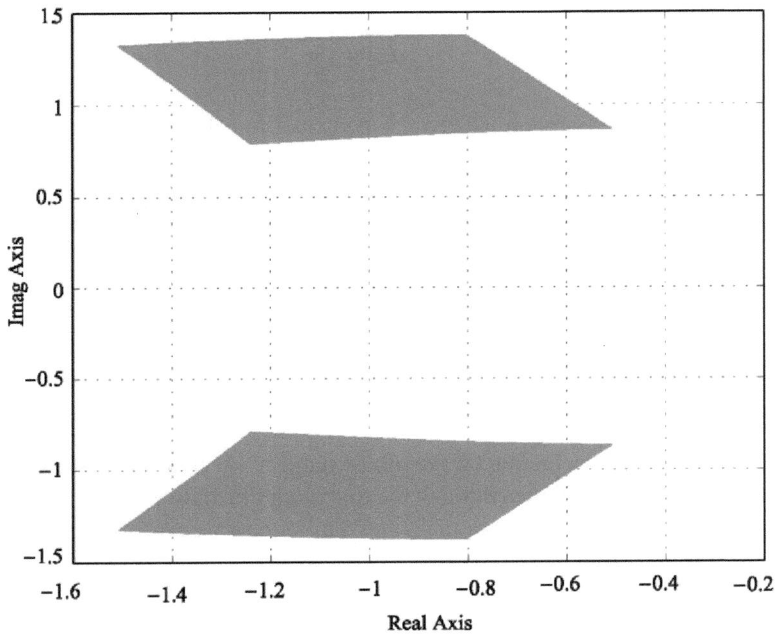

Fig. 4.14 Poles map in s-complex plane for LTI FOS (4.49) with parametric interval uncertainties $a \in [0,1]$ and $b \in [0,1]$.

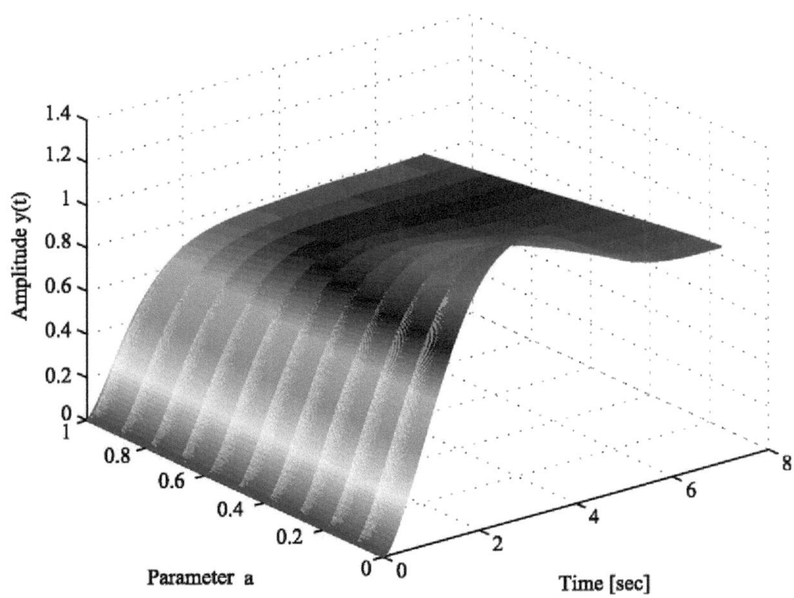

Fig. 4.15 Transient unit step responses to (4.49) for variable parameter $a \in [0,1]$ and to fixed value of b $(b = 1)$.

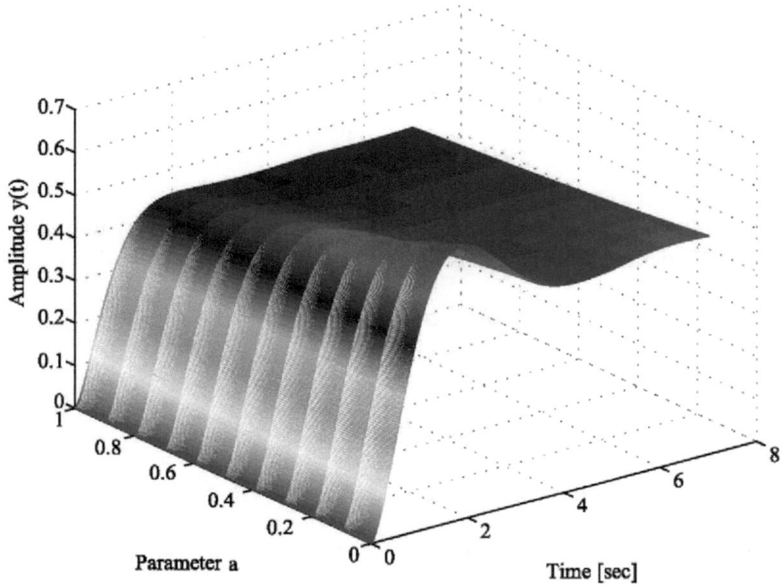

Fig. 4.16 Transient unit step responses to (4.49) for variable parameter $a \in [0,1]$ and to fixed value of b $(b = 2)$.

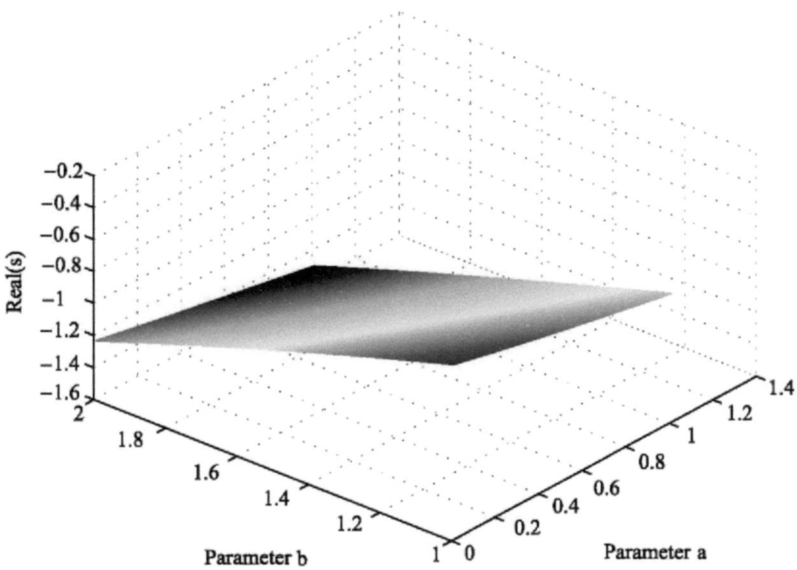

Fig. 4.17 Real parts of pseudo-polynomial poles projected in 3D view according to parameters a and b.

Kharitonov-Based Method:

Applying the robust stability theorem, however, we have the equivalence system

$$H(\sigma) = \frac{1}{\sigma^3 + a\sigma + b}, \tag{4.51}$$

where the transformation is $s^\alpha \to \sigma$, $\alpha = 0.5$. The new characteristic polynomial with uncertain parameters a and b is

$$P(\sigma) = \sigma^3 + a\sigma + b = 0.$$

An interval polynomial is then

$$p(\sigma, q) = [1,2] + [0,1]\sigma + \sigma^3$$

and four Kharitonov's polynomials are:

$$\begin{aligned} p^{--}(\sigma) &= 1 + \sigma^3, \\ p^{-+}(\sigma) &= 1 + \sigma + \sigma^3, \\ p^{+-}(\sigma) &= 2 + \sigma^3, \\ p^{++}(\sigma) &= 2 + \sigma + \sigma^3. \end{aligned} \tag{4.52}$$

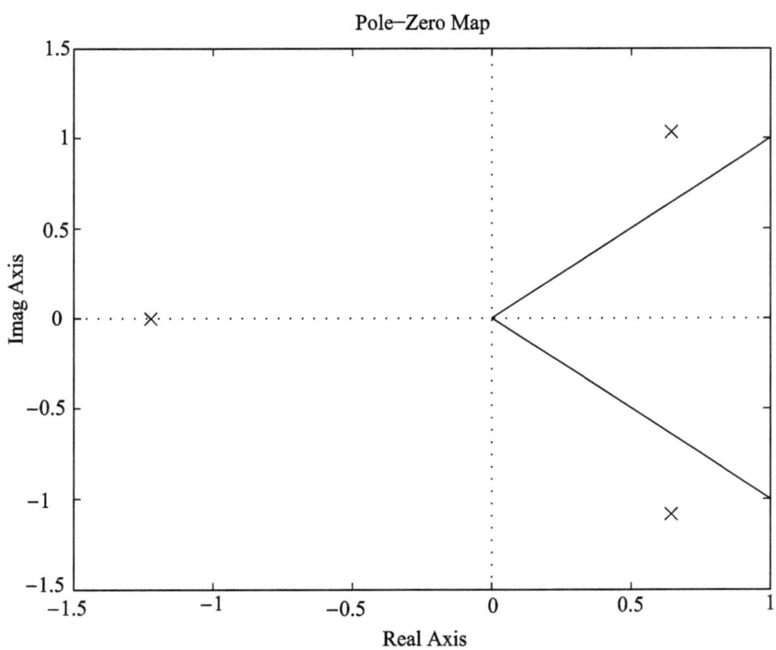

Fig. 4.18 Poles map in σ-complex plane and stability limits (closed angular sector) for $\alpha = 0.5$.

For our case, where polynomial degree is 3, it will be sufficient to test the polynomial $p^{+-}(\sigma)$. The roots of the polynomial $p^{+-}(\sigma) = 2 + \sigma^3$ are: $\sigma_{1,2} = 0.629 \pm 1.091j$, $\sigma_3 = -1.265$. Fig. 4.18 shows the poles map, where all poles lie inside the limited angular sector because $|\arg(\sigma) > \pi/4|$ for all σ, where $|\arg(\sigma_{1,2})| = 1.0472$ and $|\arg(\sigma_3)| = \pi$ for $\alpha = 0.5$.

The results given above show that system $G(s,a,b)$ is *robustly stable* for all $a \in [0,1]$ and $b \in [0,1]$.

4.4.2 Stability Check When the Fractional Orders Are Also Interval Real Numbers

In the general case of *fractional-order* systems the transfer function has the form of (3.2). Here, we consider a more specific form of (3.2) with interval uncertainties in the fractional orders and parameters. Then, the Kharitonov polynomial can be expressed in the following "new" general form:

$$P(s,o,q) = \sum_{i=0}^{n} [q_i^-, q_i^+] s_i^{[o^-,o^+]}, \tag{4.53}$$

where $q^{\pm}$ are parametric interval uncertainties and $o^{\pm}$ are interval order uncertainties.

Now the question is: Can we extend the Kharitonov theorem to interval polynomial with interval orders as in (4.53) as well? Our conjecture is "yes" for some situations. Intuitively, we should have 16 edge polynomials: $p_{--}^{--}(s)$, $p_{-+}^{--}(s)$, $p_{+-}^{--}(s)$, $p_{++}^{--}(s)$, $p_{--}^{-+}(s)$, $p_{-+}^{-+}(s)$, $p_{+-}^{-+}(s)$, $p_{++}^{-+}(s)$, $p_{--}^{+-}(s)$, $p_{-+}^{+-}(s)$, $p_{+-}^{+-}(s)$, $p_{++}^{+-}(s)$, $p_{--}^{++}(s)$, $p_{-+}^{++}(s)$, $p_{+-}^{++}(s)$, $p_{++}^{++}(s)$, where the upper index is for the order intervals and the lower index is for parameter intervals.

The general form above is not very appropriate for stability investigation because there is a large number of polynomials and we need to reduce this number in practice. Let us consider the combination approach. For example, if we have two intervals (one for parameter and one for order), we can write four new Kharitonov polynomials; if we have three intervals (two for parameters and one for order), we can write eight polynomials, etc. In general, we can write 2^n polynomials where n is the total number of intervals in the polynomial (4.53).

Fortunately, in practice, we usually use two or three terms at most in the fractional-order mathematical models (Podlubny, 1999). In this case, we have to investigate four or eight edge polynomials.

In what follows, we consider a special case where the noninteger-order intervals are assumed to have commensurate limits for orders. This assumption is not unreasonable in the sense that we can slightly fine adjust (enlarge) the order intervals so that the bounds are commensurate. Then, we can apply the transform method $s^\alpha \to \sigma$ to checking the stability in σ-complex plane.

Therefore, the *robust stability* test procedure for the LTI FOS of commensurate orders with parametric interval and order uncertainties is similar to the one presented in Section 4.4.1 for crisp orders. The robust stability test procedures are summarized in the following:

- **step 1:** Rewrite the LTI FOS $G(s)$ of the commensurate order α to the equivalence system $H(\sigma)$, where the transformation is: $s^\alpha \to \sigma$, $\alpha \in \mathrm{R}^+$;
- **step 2:** Write the interval polynomial $P(\sigma,o,q)$ of the equivalence system $H(\sigma)$, where interval characteristic polynomial is defined as

$$P(\sigma,o,q) = \sum_{i=0}^{n} [q_i^-,q_i^+]\sigma_i^{[o^-,o^+]};$$

- **step 3:** For interval polynomial $P(\sigma,o,q)$, construct 2^n Kharitonov polynomials;
- **step 4:** Test the 2^n Kharitonov's polynomials to clarify whether they satisfy the stability condition: $|\arg(\sigma_i)| > \alpha\frac{\pi}{2}$, $\forall\sigma \in \mathrm{C}$, with σ_i being the i-th root of $P(\sigma)$;

The *robust stability* test procedure for the LTI of fractional orders with parametric interval and order uncertainties, in general form of (3.2), can be divided into the following steps:

- **step 1:** Write the interval polynomial $P(s,o,q)$ of the system $G(s)$, where interval polynomial is defined as (4.53);
- **step 2:** For interval polynomial $P(s,o,q)$, construct 2^n Kharitonov polynomials;
- **step 3:** Test the 2^n Kharitonov's polynomials to clarify whether they satisfy the stability condition: $|\arg(s_i)| > \frac{\pi}{2}$, $\forall s \in \mathrm{C}$, with s_i being the i-th root of $P(s)$.

Note that for pseudo-polynomials of fractional power we can effectively use Matlab Symbolic Math Toolbox. We will demonstrate this technique for the robust stability check upon the LTI FOS with parametric and order interval uncertainties through the following illustrative examples.

Example 4.12. : Consider the simple case of the fractional-order system, where fractional order α is within the interval $\alpha \in [1,2)$:

$$G(s) = \frac{1}{s^\alpha + 1}. \tag{4.54}$$

We assume that the above simple plant will stabilize with simple proportional controller with the transfer function $C(s) = K_p$. To investigate the stability property it is necessary to check the following characteristic pseudo-polynomial:

$$p(s) = s^\alpha + 1 + K_p, \tag{4.55}$$

where K_p is the controller gain. In this particular case we will use the controller gain $K_p = 2$ and then we obtain the characteristic pseudo-polynomial in the form $p(s) = s^\alpha + 3$, where α is within the considered interval $\alpha \in [1,2)$.

The resulting transfer function of closed control loop has the following form:

$$G_c(s) = \frac{2}{s^{\alpha}+3}. \tag{4.56}$$

Graphical Method:

Figure 4.19 shows the pole positions in the complex plane for characteristic pseudo-polynomial with various values of α.

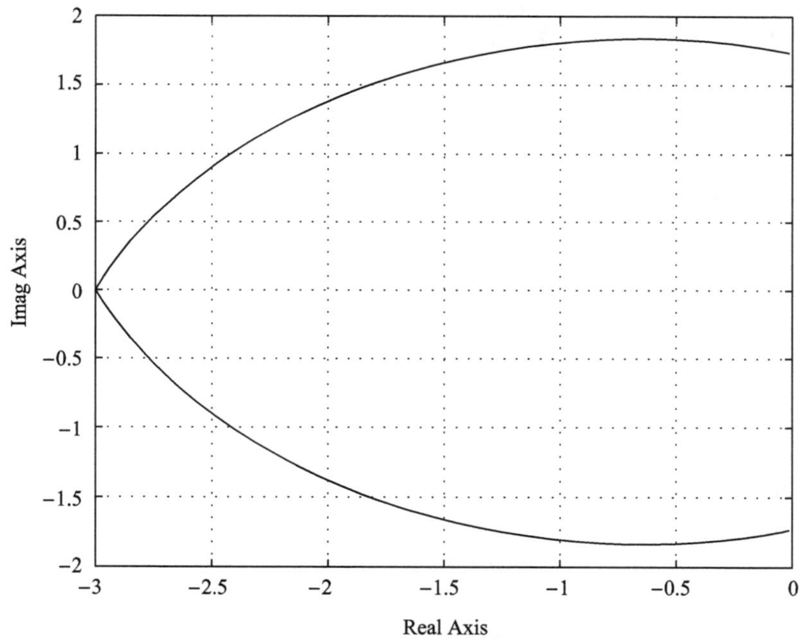

Fig. 4.19 Pole locations for characteristic pseudo-polynomial $p(s) = s^{\alpha}+3$ for $\alpha \in [1,2)$.

Figure 4.20 shows the unit step responses to various values of order $\alpha \in [1,2)$ and to time $T_{sim} = 7\,s$.

Remark 4.2. If we compare the unit step responses depicted in Fig.4.20 and the pole distribution depicted in Fig.4.19, we can see the correlation. For the single pole $[-3]$ case when $\alpha = 1$ (lower limit of α), a monotonous unit step response results while for the complex conjugate roots $[\pm 1.73]$ case when $\alpha = 2$ (upper limit of α), an oscillating unit step response results.

Consider the other unknown parameter a which is assumed to lie in a known interval $a \in [1,2]$. The transfer function of controlled system is

$$G(s) = \frac{1}{s^{\alpha}+a} \tag{4.57}$$

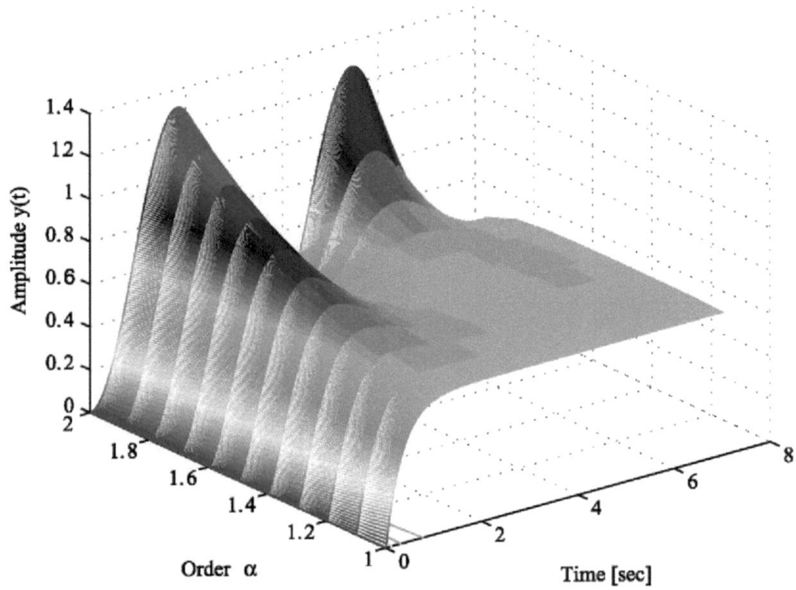

Fig. 4.20 Transient unit step responses to (4.56) for variable order $\alpha \in [1,2)$.

and the characteristic pseudo-polynomial of closed control loop has the form $p(s) = s^{\alpha} + a + 2$.

The resulting closed-loop transfer function has the following form:

$$G_c(s) = \frac{2}{s^{\alpha} + a + 2}. \tag{4.58}$$

The unit-step response was derived from (3.9) and has the form

$$y(t) = 2\mathscr{E}_0(t, -(a+2); \alpha, \alpha+1) \tag{4.59}$$

with zero initial conditions.

The pole locations in the complex plane for the characteristic pseudo-polynomial with various values of orders α and a are shown in Fig. 4.21 where we can see that the poles in the stability region of the complex plane have been changed from curves to a plane, which depends on the values α and a in the pseudo-polynomial.

Figure 4.22 shows the unit step time responses to various values of the order $\alpha \in [1,2)$ with a fixed $a = 2$.

Remark 4.3. Comparing the unit step responses in Fig. 4.20 and the unit step responses in Fig. 4.22, we can see another correlation. Both responses are for variable values of order $\alpha \in [1,2)$, But the first is for the lower limit of parameter $a = 1$ and the second is for the upper limit of parameter $a = 2$. The difference correlates with the poles distribution depicted in Fig. 4.21. With $a = 2$, we have a single pole $[-4]$. When $\alpha = 1$ (lower limit of α), we obtain a monotone unit step response. Similarly,

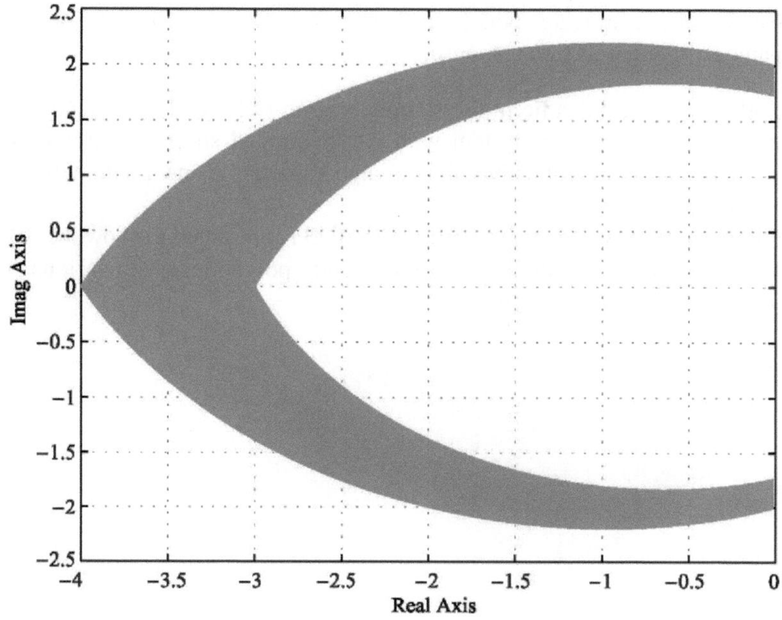

Fig. 4.21 Poles location for characteristic pseudo-polynomial $p(s) = s^{\alpha} + a + 2$ for $\alpha \in [1,2)$ and $a \in [1,2]$.

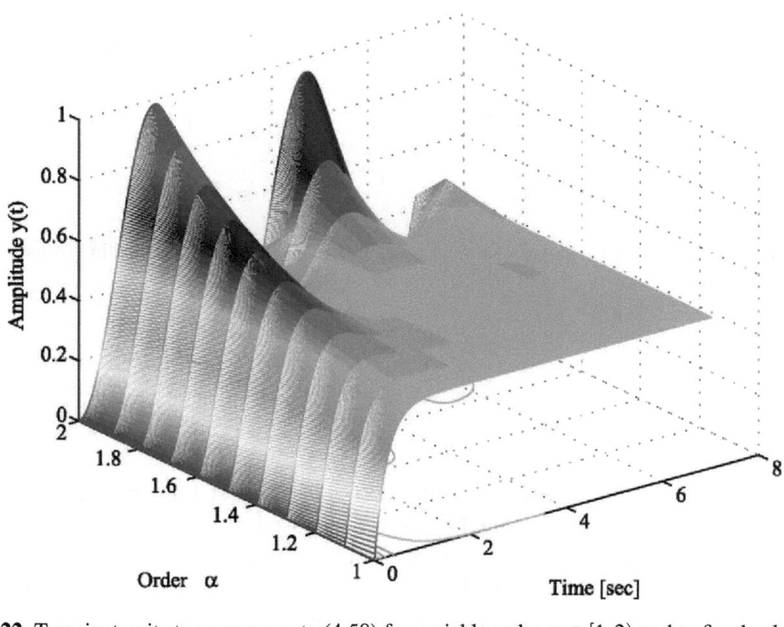

Fig. 4.22 Transient unit step responses to (4.58) for variable order $\alpha \in [1,2)$ and to fixed value of a $(a = 2)$.

we have complex conjugate roots [± 2.0] when $\alpha = 2$ (upper limit of α) and we obtain an oscillating unit step response.

In both cases, the poles lie in the stability region (left-hand half-plane) and we can say that the closed-loop system with the mentioned simple plant and simple proportional controller is robustly stable for all parameters α and a within the known considered intervals.

For higher visibility of robust stability, Fig. 4.23 is presented to show the 3D projection of the stable plane for characteristic pseudo-polynomial with various values of the order α and the coefficient a.

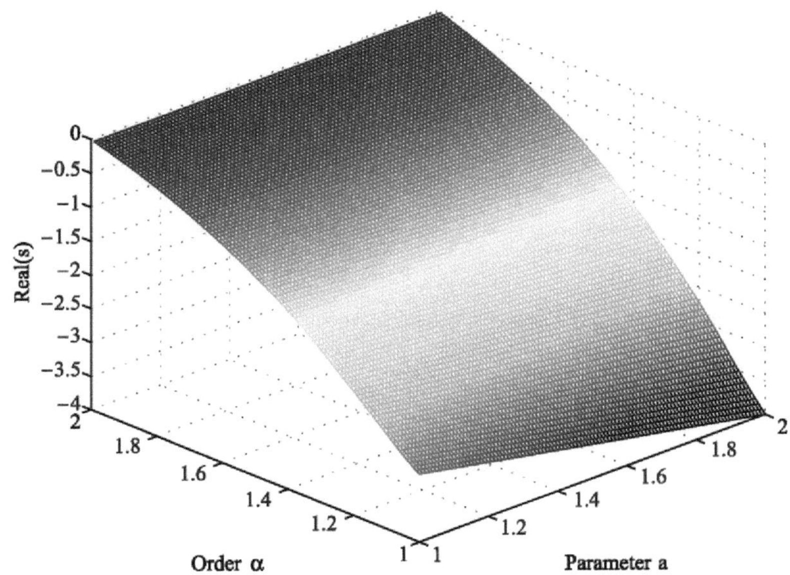

Fig. 4.23 Real parts of pseudo-polynomial poles projected in 3D view according to parameter a and real order α.

Kharitonov-Based Method:

The graphical method presented in the above is the best for the validation of the Kharitonov-based stability check procedure proposed in Subsection 4.4.2.

From Fig. 4.21, we can observe the limit points for the bounded shape in the complex plane: [-3, -4] on the real axis and [$\pm 1.73j, \pm 2j$] on the imaginary axis.

If we return to characteristic pseudo-polynomial $p(s) = s^{\alpha} + a + K$ for $\alpha \in [1,2)$ and $a \in [1,2]$, we can write the following "new" form for polynomial (for $K = 2$):

$$p(s) = s^{[1,2]} + [1,2] + 2. \tag{4.60}$$

We assume that such forms of polynomials are continuous and analytical functions and have roots, which can be single or complex-conjugate (branch points). We can write the following polynomials based on Kharitonov's principle but modified for interval orders as well:

$$
\begin{aligned}
{}^1p(s) &= s+1+2 = s+3 \rightarrow s = -3,\\
{}^2p(s) &= s+2+2 = s+4 \rightarrow s = -4,\\
{}^3p(s) &= s^2+1+2 = s^2+3 \rightarrow s_{1,2} = \pm 1.73j,\\
{}^4p(s) &= s^2+2+2 = s^2+4 \rightarrow s_{1,2} = \pm 2.
\end{aligned} \tag{4.61}
$$

From the polynomials in (4.61), we actually can obtain the limit points for a shape in the complex plane as depicted in Fig. 4.21, as roots of each polynomial, respectively.

When the order interval limits are commensurate, we can use the method described in Section 4.4.2. The characteristic pseudo-polynomial (4.60) can be rewritten by using the transform method $s^\alpha \rightarrow \sigma$ as a polynomial in the following form:

$$p(\sigma) = \sigma^{[2,4]} + [1,2] + 2, \tag{4.62}$$

where the transformation is such that $s^{0.5} \rightarrow \sigma$.

According to the proposed test procedure, we can write the following four polynomials:

$$
\begin{aligned}
{}^1p(\sigma) &= \sigma^2+3 \rightarrow \sigma_{1,2} = \pm 1.732j,\\
{}^2p(\sigma) &= \sigma^2+4 \rightarrow \sigma_{1,2} = \pm 2.000j,\\
{}^3p(\sigma) &= \sigma^4+3 \rightarrow \sigma_{1,2,3,4} = \pm 0.930 \pm 0.930j,\\
{}^4p(\sigma) &= \sigma^4+4 \rightarrow \sigma_{1,2,3,4} = \pm 1.000 \pm 1.000j.
\end{aligned} \tag{4.63}
$$

Figure 4.24 shows the pole locations in σ- complex plane for the characteristic polynomials (4.63). The closed angular stability limits for $\alpha\pi/2$, for $\alpha = 0.5$, are also drawn in Fig. 4.24.

The system is stable because all roots $\sigma_i, \forall i$ of polynomials (4.63) have $|\arg(\sigma_i)| > \alpha\frac{\pi}{2}$, for $\alpha = 0.5$.

This result validates the results depicted in Fig. 4.21 and Fig. 4.23. Based on these results, we can declare that the system (4.57) is robustly stable for interval parametric uncertainties and interval order uncertainties and can be controlled by a simple proportional controller $C(s) = 2$.

Example 4.13. : Let us recall the application of Example 4.5 already considered in this chapter for stability investigation and analysis, which is based on the work of Petráš et al. (2002b), where the application of fractional-order controller for temperature control of heat solid (electrical heater) was performed.

To obtain mathematical models of the controlled object, various identification methods can be used. In (Petráš et al., 2002b), two models were chosen as the nom-

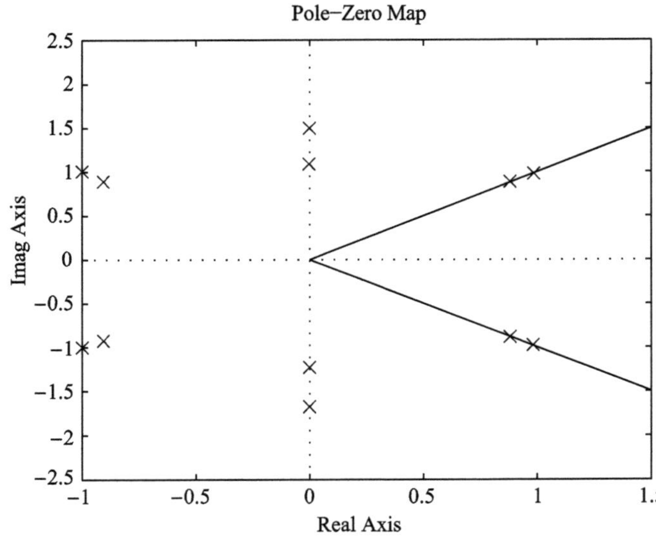

Fig. 4.24 Pole locations for the polynomials (4.63) and the stability limits $\pi/4$.

inal models for control system design. The first one is an integer first-order model given by

$$G_I(s) = \frac{1}{20.14s + 0.598}. \tag{4.64}$$

The second one is a fractional-order one-term model of the following form:

$$G_F(s) = \frac{1}{39.69s^{1.25} + 0.598}. \tag{4.65}$$

The second one is closer to the reality based on our experience. For comparison, the integer first-order model is also used (Petráš et al., 2002b) with the standard control design techniques. However, the reality is probably somewhere *in between* the $G_I(s)$ and $G_F(s)$.

Now, we follow a new idea of using the standard control design method with a *PD* controller (particular case of PD^δ controller for $\delta = 1$) for control of this object but two assumed uncertainties in the mathematical model of the controlled object. The new uncertain model has the following form:

$$G(s) = \frac{1}{as^\alpha + 0.598}, \tag{4.66}$$

where $a \in [15, 45]$ and $\alpha \in [1, 1.3]$.

With the *PD* controller

$$C(s) = K + T_d s,$$

to achieve the desired stability measure $S_t = 2.0$, the following parameters were designed (Petráš et al., 2002b): $K = 64.47$ and $T_d = 12.46$, which means that the closed-loop characteristic equation has a simple root [-2].

Now, we apply the designed *PD* controller $C(s) = 64.47 + 12.46s$ to the new uncertain model (4.66). The resulting closed-loop transfer function becomes

$$G_c(s) = \frac{12.46s + 64.47}{as^{\alpha} + 12.46s + 65.068}, \tag{4.67}$$

where $a \in [15, 45]$ and $\alpha \in [1, 1.3]$.

The unit-step response of system (4.67) is obtained by (3.9) as follows:

$$\begin{aligned} y(t) = {} & \frac{12.46}{a} \sum_{k=0}^{\infty} \frac{(-1)^k}{k!} \left(\frac{12.46}{a}\right)^k \times \mathscr{E}_k(t, -\frac{65.068}{a}; \alpha, \alpha - k) \\ & + \frac{64.47}{a} \sum_{k=0}^{\infty} \frac{(-1)^k}{k!} \left(\frac{65.068}{a}\right)^k \times \mathscr{E}_k(t, -\frac{12.46}{a}; \alpha - 1, \alpha + k + 1), \end{aligned}$$

with zero initial conditions.

The characteristic polynomial of the control system (4.27) with reference to intervals for order α and parameter a has the following form:

$$p(s) = [15, 45]s^{[1, 1.3]} + 12.46s + 65.068 = 0. \tag{4.68}$$

According to the stability test procedure in Section 4.4.2, we should check the following four polynomials:

$$\begin{aligned} {}^1p(s) &= 27.46s + 65.068 \rightarrow s = -2.37, \\ {}^2p(s) &= 15s^{1.3} + 12.46s + 65.068, \rightarrow s_{1,2} = -1.97 \pm 1.04j, \\ {}^3p(s) &= 57.46s + 65.068 \rightarrow s = -1.13, \\ {}^4p(s) &= 45s^{1.3} + 12.46s + 65.068 \rightarrow s_{1,2} = -0.94 \pm 0.65j. \end{aligned} \tag{4.69}$$

From the polynomials in (4.69), we can obtain the limit points for the shape of the pole locations in the complex plane as the roots of each polynomial. Clearly, the shape is situated in the left-hand side of the complex plane which means that the closed-loop system is stable. According to the proposed test procedure, the system is stable as well because all roots s_i, $\forall i$ of polynomials (4.69) have $|\arg(s_i)| > \frac{\pi}{2}$.

Based on these results, we can declare that system (4.66) is robustly stable for interval parametric and fractional order uncertainties and can be controlled by the *PD* controller $C(s) = 64.47 + 12.46s$.

Figure 4.25 shows the unit step responses for various values of order $\alpha \in [1, 1.3]$ with a fixed value of variable a $(a = 15)$.

Similarly, Fig. 4.26 shows the unit step responses to various values of order $\alpha \in [1, 1.3]$ with a fixed value of variable a $(a = 45)$.

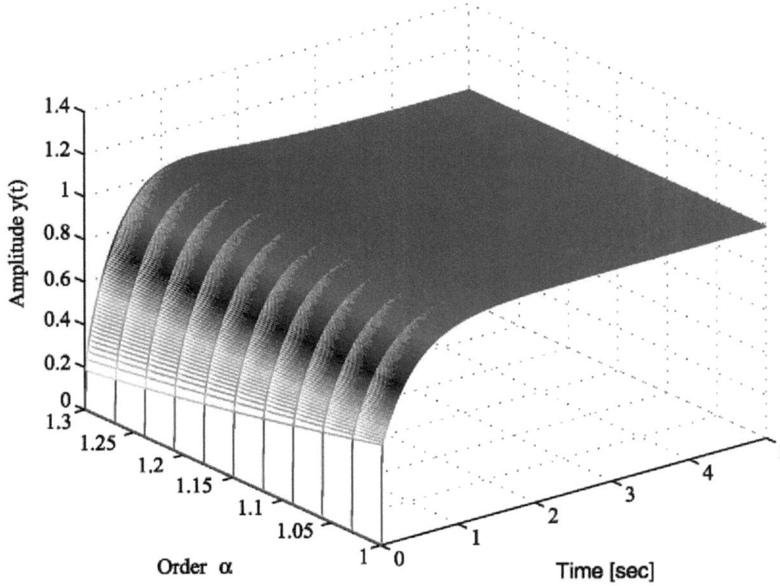

Fig. 4.25 Transient unit step responses to (4.27) for variable order $\alpha \in [1, 1.3]$ and to fixed value of a $(a = 15)$.

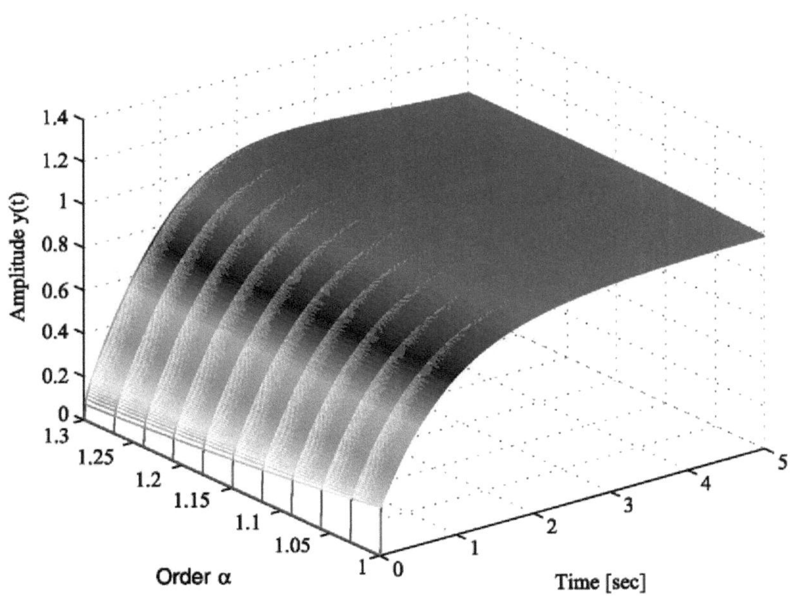

Fig. 4.26 Transient unit step responses to (4.27) for various order $\alpha \in [1, 1.3]$ and to fixed value of a $(a = 45)$.

4.5 Stability of Fractional-Order Nonlinear Uncertain Systems

As we mentioned in previous sections, there are several methods to study stability of linear fractional-order linear systems with uncertainties. However, for nonlinear fractional-order systems, there is a lack of tools for performing the stability investigation.

The stability analysis of autonomous nonlinear fractional-order systems can be converted to a problem of investigating the existence of oscillations known as limit cycles. If the limit cycles are predicted, then it is also interesting to know the number of limit cycles, their frequencies, amplitudes and other characteristics.

In paper by Nataraj and Kalla (2009) was proposed an algorithm to compute limit cycles for the uncertain nonlinear fractional-order system with a prescribed accuracy. The uncertain parameter may be associated with either the linear or nonlinear element in the system.

Consider the following uncertain fractional-order system with parametrical uncertainties (N'Doye et al., 2009):

$$\begin{aligned} {}_0D_t^{\mathbf{q}}\mathbf{x}(t) &= A\mathbf{x}(t) + B\mathbf{u}(t) + \Delta A(x) + \Delta B(u), \\ y(t) &= C\mathbf{x}(t) + D\mathbf{u}(t) + \Delta C(x) + \Delta D(u), \\ x^{(k)}(0) &= x_0^{(k)}, \quad k = 0, 1, \ldots, n-1, \end{aligned} \tag{4.70}$$

where $\Delta A(x)$, $\Delta B(u)$, $\Delta C(x)$, and $\Delta D(u)$ are nonlinear parametrical uncertainties. Suppose that the pair (A,B) is controllable and the pair (C,A) is the observable matrix.

If the uncertainties are unknown but can be decribed by the input-output form, then

$$\begin{aligned} {}_0D_t^{\mathbf{q}}\mathbf{x}(t) &= A\mathbf{x}(t) + B\mathbf{u}(t), \\ y(t) &= C\mathbf{x}(t) + D\mathbf{u}(t) + \Delta h(u), \\ x^{(k)}(0) &= x_0^{(k)}, \quad k = 0, 1, \ldots, n-1, \end{aligned} \tag{4.71}$$

where the model uncertainty $\Delta h(u)$ is nonlinear and bounded $|\Delta h(u)| \leq c|u|$, where c is positive constant.

Consider the Lur'e problem, which has a forward path that is linear and time-invariant and a feedback path that contains memoryless, possibly time-varying, nonlinear function which is piecewise continuous in t. The linear part can be characterized by four matrices (A,B,C,D), in the case of state space representation or transfer function $L(s)$, while the nonlinear part is $h(u)$ with $h(u)/u \in [a,b]$, $a < b$, $\forall y$ and the nonlinear element $h(u)$ is a time-invariant nonlinearity belonging to open sector $(0,\infty)$. The Lur'e problem is to derive conditions involving only the transfer matrix $L(s)$ and a, b so that $x = 0$ is a globally uniformly asymptotically stable equilibrium of the system. From the view of modern robustness theory, the absolute stability theory can be considered as the first approach to robust stability of nonlinear uncertain systems. An illustrative example of the uncertain neutral type Lur'e system and robust absolute stability criteria for this system can be found (Han et al., 2008).

In Fig. 4.27 is presented a basic feedback system, which is the so-called Lur'e problem. There exist two main theorems concerning this problem: the circle criterion and the Popov criterion, which is applicable only to autonomous systems.

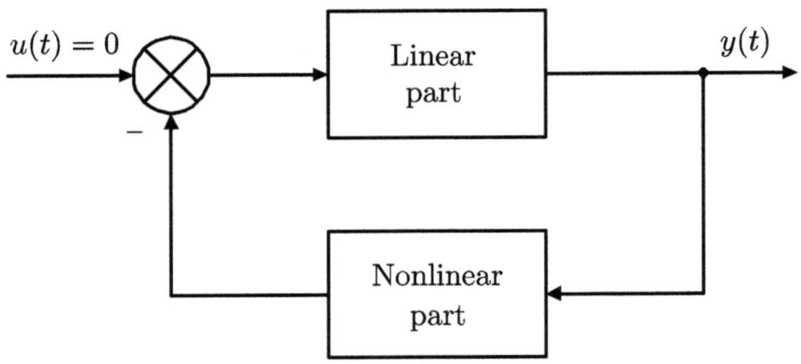

Fig. 4.27 Lur'e problem diagram.

Let us consider the following general autonomous fractional-order nonlinear system with uncertainties:

$$\begin{aligned} {}_0D_t^{\mathbf{q}}\mathbf{x}(t) &= \mathbf{f}(t,\mathbf{x}(t)) + \Delta\mathbf{f}(\mathbf{x}), \\ x^{(k)}(0) &= x_0^{(k)}, \quad k = 0,1,\ldots,n-1, \end{aligned} \tag{4.72}$$

where ${}_0D_t^q$ is the Caputo's fractional derivative of order q, and $\Delta\mathbf{f}(\mathbf{x})$ is vector of uncertain parameters, which is ussualy bounded $|\Delta\mathbf{f}(\mathbf{x})| \leq c$.

When the fractional-order nonlinear system with uncertainties has a sector nonlinearity it can be reduced to Lur'e problem and solved via method described (Han et al., 2008). It is possible only for very limited class of nonlinear systems. In some special cases of the nonlinear system with uncertanties in the linear part of the system we may use the techniques described in previous section. However, in the case of system with uncertainties in nonlinear part it is not possible. In general, the robust stability analysis of the uncertain nonlinear fractional-order system is an open problem.

References

Ahmed E., El-Sayed A. M. A. and El-Saka H. A. A., 2006, On some Routh-Hurwitz conditions for fractional order differential equations and their applications in Lorenz, Rössler, Chua and Chen systems, *Physics Letters A*, **358**, 1–4.

Ahn H. S. and Chen Y. Q., 2008, Necessary and suficient stability condition of fractional-order interval linear systems, *Automatica*, **44**, 2985–2988.

Akcay H. and Malti R., 2008, On the completeness problem for fractional rationals with incommensurable differentiation orders, *Proc. of the 17th World Congress IFAC*, Soul, Korea, July 6–11, 15367–15371.

Anderson B. D. O., Bose N. I. and Jury E. I., 1974, A simple test for zeros of a complex polynomial in a sector, *IEEE Transactions on Automatic Control*, **AC-19**, 437–438.

Aoun M., Malti R., Levron F. and Oustaloup A., 2004, Numerical simulations of fractional systems: An overview of existing methods and improvements, *Nonlinear Dynamics*, **38**, 117–131.

Aoun M., Malti R., Levron F. and Oustaloup A., 2007, Synthesis of fractional Laguerre basis for system approximation, *Automatica*, **43**, 1640–1648.

Bayat F. M. and Afshar M., 2008, Extending the root-locus method to fractional-order systems, *Journal of Applied Mathematics*, Article ID **528934**.

Bayat F. M. and Ghartemani M. K., 2008, On the essential instabilities caused by multi-valued transfer functions, *Math. Problems in Engineering*, Article ID **419046**.

Bayat F. M., Afshar M. and Ghartemani M. K., 2009, Extension of the root-locus method to a certain class of fractional-order systems, *ISA Transactions*, **48**, 48–53.

Bellman R., 1953, *Stability Theory of Differential Equations*, McGraw-Hill Book Company, New York.

Bloch F., 1946, Nuclear induction, *Physics Review*, **70**, 460–473.

Bonnet C. and Partington J. R., 2000, Coprime factorizations and stability of fractional differential systems, *Systems and Control Letters*, **41**, 167–174.

Charef A., 2006, Modeling and analog realization of the fundamental linear fractional order differential equation, *Nonlinear Dynamics*, **46**, 195–210.

Chen Y. Q. and Moore K. L., 2002, Analytical stability bound for a class of delayed fractional-order dynamic systems, *Nonlinear Dynamics*, **29**, 191–200.

Chen Y. Q., Ahn H. S. and Xue D., 2006a, Robust controllability of interval fractional order linear time invariant systems, *Signal Processing*, **86**, 2794–2802.

Chen Y. Q., Ahn H. S. and Podlubny I., 2006b, Robust stability check of fractional order linear time invariant systems with interval uncertainties, *Signal Processing*, **86**, 2611–2618.

Deng W., Li C. and Lu J., 2007, Stability analysis of linear fractional differential system with multiple time delays, *Nonlinear Dynamics*, **48**, 409–416.

Dorčák Ľ., 1994, Numerical Models for the Simulation of the Fractional-Order Control Systems, *UEF-04-94, The Academy of Sciences, Inst. of Experimental Physic*, Košice, Slovakia.

Dorf R. C. and Bishop R. H., 1990, *Modern Control Systems*, Addison-Wesley, New York.

Dzieliski A. and Sierociuk D., 2008, Stability of discrete fractional order state-space systems, *Journal of Vibration and Control*, **14**, 1543–1556.

El-Salam S. A. A. and El-Sayed A. M. A., 2007, On the stability of some fractional-order non-autonomous systems, *Electr. J. of Qualitative Theory of Diff. Equations*, **6**, 1–14.

Ghartemani M. K. and Bayat F. M., 2008, Necessary and sufficient conditions for perfect command following and disturbance rejection in fractional order systems, *Proc. of the 17th World Congress IFAC*, Soul, Korea, July 6–11, 364–369.

Gross B. and Braga E. P., 1961, *Singularities of Linear System Functions*, Elsevier Publishing, New York.

Haacke E. M., Brown R. W., Thompson M. R. and Venkatesan R., 1999, *Magnetic Resonance Imaging: Physical Principles and Sequence Design*, Wiley, New York.

Hamamci S. E., 2008, Stabilization using fractional-order PI and PID controllers, *Nonlinear Dynamics*, **51**, 329–343.

Han Q. L., Xue A., Liu S. and Yu X., 2008, Robust absolute stability criteria for uncertain Lure systems of neutral type, *Int. J. Robust Nonlinear Control*, **18**, 278–295.

Henrion D., 2001, *Robust stability analysis: Interval uncertainty, Graduate course on polynomial methods for robust control - Part 1 and Part 2*, http://www.laas.fr/ henrion/courses/.

Hwang C. and Cheng Y. C., 2006, A numerical algorithm for stability testing of fractional delay systems, *Automatica*, **42**, 825–831.

Kheirizad I., Tavazoei M. S. and Jalali A. A., 2009, Stability criteria for a class of fractional order systems, *Nonlinear Dyn.*, DOI: 10.1007/s11071-009-9638-1.

Kotek Z., Kubik S. and Razim M., 1973, *Nonlinear dynamical systems*, TKI-SNTL, Praha (in Czech).

LePage W. R., 1961, *Complex variables and the Laplace transform for engineers*, McGraw-Hill, New York.

Li Y., Chen Y. Q., Podlubny I. and Cao Y., 2008, Mittag-Leffler stability of fractional order nonlinear dynamic system, *Proc. of the 3rd IFAC Workshop on Fractional Differentiation and its Applications*, November 05–07, Ankara, Turkey.

Lu J. G. and Chen G., 2009, Robust stability and stabilization of fractional order interval systems: An LMI approach, *IEEE Transactions on Automatic Control*, **54**, 1294–1299.

Magin R. L., Feng X. and Baleanu D., 2009, Solving the fractional order Bloch equation, *Concepts in Magnetic Resonance Part A*, **34A**, 16–23.

Matignon D., 1996, Stability result on fractional differential equations with applications to control processing, *IMACS-SMC Proceedings*, Lille, France, July, 963–968.

Matignon D., 1998, Stability properties for generalized fractional differential systems, *Proc. of Fractional Differential Systems: Models, Methods and App.*, **5**, 145–158.

Nataraj P. S. V. and Kalla R., 2009, Computation of limit cycles for uncertain nonlinear fractional-order systems, *Physica Scripta*, Article ID **014021**.

N'Doye I., Zasadzinski M., Radhy N. E. and Bouaziz A., 2009, Robust controller design for linear fractional-order systems with nonlinear time-varying model uncertainties, *Proc. of the 17th Mediterranean Conference on Control and Automation*, June 24–26, Thessaloniki, Greece, 821–826.

Oustaloup A., Sabatier J., Lanusse P., Malti R., Melchior P., Moreau X. and Moze M., 2008, An overview of the CRONE approach in system analysis, modeling and identification, observation and control, *Proc. of the 17th World Congress IFAC*, Soul, Korea, July 6–11, 14254–14265.

Özturk N. and Uraz A., 1985, An analysis stability test for a certain class of distributed parameter systems with delay, *IEEE Transactions on Circuit and Systems*, **CAS-32**, 393–396.

Petráš I. and Dorčák Ľ., 1999, The frequency method for stability investigation of fractional control systems, *Journal of SACTA*, **2**, 75–85.

Petráš I., Chen Y. Q. and Vinagre B. M., 2002a, A robust stability test procedure for a class of uncertain LTI fractional order systems, *Proc. of the Int. Carpathian Control Conf.*, May 27–30, Malenovice, Czech republic, 247–252.

Petráš I., Vinagre B. M., Dorčák Ľ. and Feliu V., 2002b, Fractional Digital Control of a Heat Solid: Experimental Results, *Proc. of the Int. Carpathian Control Conf.*, May 27–30, Malenovice, Czech republic, 365–370.

Petráš I., Chen Y. Q. and Vinagre B. M., 2004a, Robust stability test for interval fractional order linear systems, Volume 208–210, Chapter 6.5: V. D. Blondel and A. Megretski (Eds.), *Unsolved Problems in the Mathematics of Systems and Control*, USA, Princeton University Press.

Petráš I., Chen Y. Q., Vinagre B. M. and Podlubny I., 2004b, Stability of linear time invariant systems with interval fractional orders and interval coefficients, *Proc. of the Int. Conf. on Computation Cybernetics*, 8/30–9/1, Vienna, Austria, 1–4.

Petráš I., 2009, Stability of fractional-order systems with rational orders: A survey, *Fractional Calculus and Applied Analysis*, **12**, 269–298.

Podlubny I., 1999, *Fractional Differential Equations*, Academic Press, San Diego.

Radwan A. G., Soliman A. M., Elwakil A. S. and Sedeek A., 2009, On the stability of linear systems with fractional-order elements, *Chaos, Solitons and Fractals*, **40**, 2317–2328.

Tan N., Özgüven Ö. F. and Özyetkin M. M., 2009, Robust stability analysis of fractional order interval polynomials, *ISA Transactions*, **48**, 166–172.

Tavazoei M. S. and Haeri M., 2007a, Unreliability of frequency-domain approximation in recognising chaos in fractional-order systems, *IET Signal Proc.*, **1**, 171–181.

Tavazoei M. S. and Haeri, M., 2007b, A necessary condition for double scroll attractor existence in fractional-order systems, *Physics Letters A*, **367**, 102–113.

Tavazoei M. S. and Haeri M., 2008a, Limitations of frequency domain approximation for detecting chaos in fractional order systems, *Nonlinear Analysis*, **69**, 1299–1320.

Tavazoei M. S. and Haeri M., 2008b, Chaotic attractors in incommensurate fractional order systems, *Physica D*, **237**, 2628–2637.

Tavazoei M. S. and Haeri M., 2009, A note on the stability of fractional order systems, *Mathematics and Computers in Simulation*, **79**, 1566–1576.

Vinagre B. M. and Feliu V., 2007, Optimal fractional controllers for rational order systems: A special case of the Wiener-Hopf spectral factorization method, *IEEE Transactions on Automatic Control*, **52**, 2385–389.

Yeung K. S. and Wang S. S., 1987, A Simple Proof of Kharitonovs Theorem, *IEEE Transactions on Automatic Control*, **AC-32**, 822–823.

Chapter 5
Fractional-Order Chaotic Systems

5.1 Introduction to Chaotic Dynamics

In general, a nonlinear system is a system which is not linear, that is, a system which does not satisfy the superposition principle. In mathematics, a nonlinear system is any problem, where the variables to be solved cannot be written as a linear combination of independent components. If the equation contains a nonlinear function (power or cross product), the system is nonlinear as well. The system is nonlinear also if it has a nonlinear transfer characteristic as, for example, current-voltage characteristic of a diode. Last but not least, we should mention typical nonlinearity. The system is nonlinear if there is some typical nonlinearity as, for instance, saturation, hysteresis, etc. These characteristics are basic properties of the nonlinear systems.

Nonlinear systems are very interesting to engineers, physicists and mathematicians because most real physical systems are inherently nonlinear in nature. Nonlinear equations are difficult to be solved by analytical methods and give rise to interesting phenomena such as bifurcation and chaos. Even simple nonlinear (or piecewise linear) dynamical systems can exhibit completely a unpredictable behavior, the so-called deterministic chaos. Chaos theory has been so surprising because chaos can also be found within trivial systems. At this point we have to say that the word "chaos" is not uniquely defined. In the most used sense, chaotic dynamics are dynamics originated by regular dynamical equations with no stochastic coefficients, but at the same time, with trajectories that are similar or indistinguishable from some stoachastic processes (Zaslavsky, 2005).

There are a few definitions of the chaotic dynamics, e.g.: (i) a system with at least one positive Lyapunov exponent is chaotic; (ii) a system with positive entropy is chaotic; (iii) a system equivalent to hyperbolic or Anosov system is chaotic, etc. The common part of all definitions is the existence of local instability and divergence of initially close trajectories. At the same time, all definitions are not exactly equivalent.

In the next sections we focus on the well-known nonlinear systems, which exhibit chaos and hyperchaos (e.g., (Petráš, 2009b; Petráš et al., 2009), etc.).

5.2 Concept of Chua's Circuit

5.2.1 Classical Chua's Oscillator

Classical Chua's circuit, which is shown in Fig. 5.1, is a simple electronic circuit that exhibits nonlinear dynamical phenomena such as bifurcation and chaos. In fact, in order to exhibit chaos, an autonomous electronic circuit must satisfy some essential criteria which are necessary (not sufficient) conditions for the appearance of chaos (Kennedy, 1993a): the circuit must contain at least three energy-storage elements, at least one nonlinear element and at least one locally active resistor. The Chua's diode, being a nonlinear locally active resistor, allows Chua's circuit to satisfy the last two conditions. Chua's circuit satisfies all the above-mentioned criteria. The active resistor supplies energy to separate trajectories, the nonlinearity provides folding, and the three-dimensional state space permits persistent stretching and folding in a bounded region of the state space.

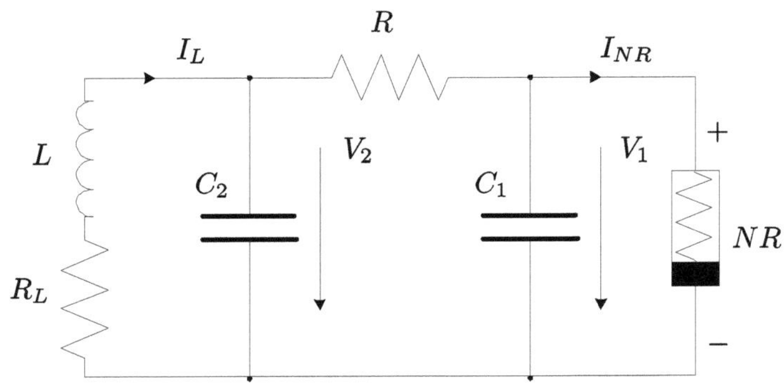

Fig. 5.1 Chua's circuit.

The simplest and most widely studied nonlinear Chua's circuit consists of five elements: two capacitors C_1 and C_2, an inductor L, a resistor R and a nonlinear resistor (NR), known as Chua's diode.

By applying Kirchhoff's circuit laws, such circuit, generally known as Chua's oscillator, can be described by the following equations (Matsumoto, 1984):

$$\begin{aligned}\frac{dV_1(t)}{dt} &= \frac{1}{C_1}\left[G(V_2(t)-V_1(t))-f(V_1(t))\right],\\ \frac{dV_2(t)}{dt} &= \frac{1}{C_2}\left[G(V_1(t)-V_2(t))+I_L(t)\right],\\ \frac{dI_L(t)}{dt} &= \frac{1}{L}\left[-V_2(t)-R_LI_L(t)\right],\end{aligned} \tag{5.1}$$

where conductance $G = 1/R$, $I_L(t)$ is the current through the inductance L, $V_1(t)$ and $V_2(t)$ are the voltages over the capacitors C_1 and C_2, respectively, and $f(V_1(t))$ is the piecewise-linear $v-i$ characteristic of NR - Chua's diode, depicted in Fig. 5.2, which can be described by the following state equations:

$$I_{NR}(t) = f(V_1(t)) = G_b V_1(t) + \frac{1}{2}(G_a - G_b)(|V_1(t) + B_p| - |V_1(t) - B_p|), \quad (5.2)$$

with B_p being the breakpoint voltage of a diode, and $G_a < 0$ and $G_b < 0$ being some appropriate constants (slope of the piecewise-linear resistance).

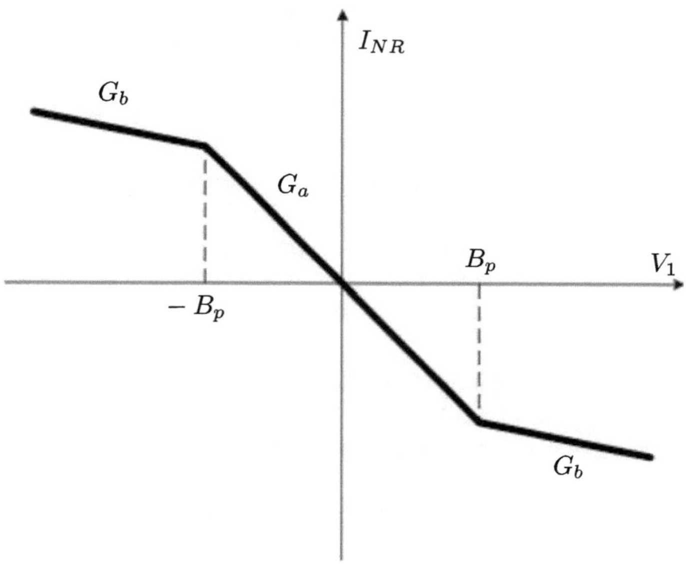

Fig. 5.2 Typical three-segment piecewise-linear $v-i$ characteristic of the nonlinear resistor.

By defining the rescaling

$$\begin{aligned} x &= V_1/B_p, \; y = V_2/B_p, \quad z = I_L/B_pG, \\ \alpha &= C_2/C_1, \; \beta = C_2/(LG^2), \gamma = C_2R/(LG), \\ m_1 &= G_b/G, \; m_0 = G_a/G, \quad \tau = t|G/C_2|, \end{aligned} \quad (5.3)$$

we can transform (5.1) into the following corresponding dimensionless form of Chua's circuit (Chua et al., 1993; Deregel, 1993):

$$\begin{aligned} \frac{dx(t)}{dt} &= \alpha\,(y(t) - x(t) - f(x)), \\ \frac{dy(t)}{dt} &= x(t) - y(t) + z(t), \end{aligned} \quad (5.4)$$

$$\frac{dz(t)}{dt} = -\beta y(t) - \gamma z(t),$$

where

$$f(x) = m_1 x(t) + \frac{1}{2}(m_0 - m_1) \times (|x(t)+1| - |x(t)-1|) \tag{5.5}$$

and τ in transformation equations (5.3) is the dimensionless time.

Because of the piecewise-linear nature of *NR*, the vector field of Chua's circuit can be decomposed into three distinct affine regions. It depends on the values of $\pm B_p$. We call these regions the outer D_{-1}, $(V_1 < -B_p)$, the inner (middle) region $D_0, (|V_1| < B_p)$ and the outer $D_1, (V_1 > B_p)$, respectively. The global dynamics of Chua's circuit may be determined by piecing together the three-dimensional vector fields of the regions D_{-1}, D_0, and D_1, then we obtain a qualitative description of the whole circuit.

The equilibrium points of Chua's oscillator (5.1) are defined by

$$\begin{aligned} 0 &= \frac{1}{C_1}\left[G(V_2(t) - V_1(t)) - I_{NR}(t))\right], \\ 0 &= \frac{1}{C_2}\left[G(V_1(t) - V_2(t)) + I_L(t)\right], \\ 0 &= \frac{1}{L}\left[-V_2(t) - R_L I_L(t)\right], \end{aligned} \tag{5.6}$$

where $I_{NR}(t)$ is given by relation (5.2). The origin is obviously an equilibrium point.

In D_0 (inner) region, where $|V_1| \le B_p$, the state equations of Chua's oscillator are linear. The Jacobian matrix has the following form:

$$\mathbf{J}_{G_a} = \begin{bmatrix} \frac{-(G+G_a)}{C_1} & \frac{G}{C_1} & 0 \\ \frac{G}{C_2} & \frac{-G}{C_2} & \frac{1}{C_2} \\ 0 & -\frac{1}{L} & -\frac{R_L}{L} \end{bmatrix}. \tag{5.7}$$

The characteristic equation of Chua's system in the inner region is

$$\begin{aligned} \det|\lambda\mathbf{I} - \mathbf{J}_{G_a}| = \lambda^3 &+ \left(\frac{G+G_a}{C_1} + \frac{G}{C_2} + \frac{R_L}{L}\right)\lambda^2 \\ &+ \left(\frac{GG_a}{C_1C_2} + \frac{G+G_a}{C_1L}R_L + \frac{GR_L}{C_2L} + \frac{1}{C_2L}\right)\lambda \\ &+ \frac{R_L GG_a + (G+G_a)}{C_1C_2L} = 0. \end{aligned} \tag{5.8}$$

In D_{-1} and D_1 (outer) regions, where $|V_1| > B_p$, the state equations of Chua's oscillator are linear. The Jacobian matrix has the following form:

$$\mathbf{J}_{G_b} = \begin{bmatrix} \frac{-(G+G_b)}{C_1} & \frac{G}{C_1} & 0 \\ \frac{G}{C_2} & \frac{-G}{C_2} & \frac{1}{C_2} \\ 0 & -\frac{1}{L} & -\frac{R_L}{L} \end{bmatrix}. \tag{5.9}$$

The characteristic equation of Chua's system in the outer region is

$$\det|\lambda \mathbf{I} - \mathbf{J}_{G_b}| = \lambda^3 + \left(\frac{G+G_b}{C_1} + \frac{G}{C_2} + \frac{R_L}{L}\right)\lambda^2 + \left(\frac{GG_b}{C_1C_2} + \frac{G+G_b}{C_1 L}R_L + \frac{GR_L}{C_2 L} + \frac{1}{C_2 L}\right)\lambda + \frac{R_L G G_b + (G+G_b)}{C_1 C_2 L} = 0. \tag{5.10}$$

The dynamic behavior of any type of Chua's circuit is determined by the six eigenvalues (Chua et al., 1986). They can be obtained by solving the characteristic equations (5.8) and (5.10) and depend on the value of electrical components.

When $G > |G_a|$ or $G < |G_b|$, the circuit has a unique equilibrium point at the origin and two virtual equilibria E_- and E_+ (lie outside regions D_{-1} and D_1). Otherwise it has three equilibrium points ar E_-, 0, and E_+. The equilibrium point E_- in the D_{-1} region has three eigenvalues. It usually consists of a real (λ_1) and a pair of complex conjugate values ($\lambda_{2,3}$). We assume that eigenvalue λ_1 is stable and eigenvalues $\lambda_{2,3}$ are unstable. With symmetry, it is the same to equilibrium point E_+.

For more detailed description of stability analysis of equilibrium points, the dynamics of the outer and inner regions, we refer the reader to (Kennedy, 1993b; Pivka et al., 1994).

Given the techniques of fractional calculus, there are a number of ways in which the order of system could be amended. In the next parts we will show several of them.

5.2.2 Fractional-Order Chua's Oscillator

As we already mentioned in Chapter 2, there are many electric and magnetic phenomena where the fractional calculus can be used. In this section we consider two of them — models of real capacitor and real inductor.

The circuit behavior can be described by three fractional differential equations with various orders. Applying Kirchhoff's laws for two current nodes and one voltage loop and relation (2.75), and (2.78) into circuit depicted in Fig. 5.1, we obtain the following mathematical model of the circuit for state variables $V_1(t)$, $V_2(t)$ and $I(t)$:

$$
\begin{aligned}
C_1\, {}_0D_t^{q_1} V_1(t) + I_{NR}(t) &= \frac{V_2(t) - V_1(t)}{R_2}, \\
C_2\, {}_0D_t^{q_2} V_2(t) - I(t) &= \frac{V_1(t) - V_2(t)}{R_2}, \\
L_1\, {}_0D_t^{q_3} I(t) + V_2(t) + R_L I(t) &= 0.
\end{aligned}
\tag{5.11}
$$

Equations (5.11) can be rewritten into the following form (Petráš, 2008):

$$
\begin{aligned}
{}_0D_t^{q_1} V_1(t) &= \frac{1}{C_1 R_2}[V_2(t) - V_1(t)] - \frac{f(V_1(t))}{C_1}, \\
{}_0D_t^{q_2} V_2(t) &= \frac{1}{C_2 R_2}[V_1(t) - V_2(t)] + \frac{I(t)}{C_2}, \\
{}_0D_t^{q_3} I(t) &= \frac{1}{L_1}[-V_2(t) - R_L I(t)],
\end{aligned}
\tag{5.12}
$$

where V_1 is the voltage across the capacitor C_1, V_2 is the voltage across the capacitor C_2, I is the current through the inductance L_1, q_1 is the real order of the capacitor C_1, q_2 is the real order of the capacitor C_2, q_3 is the real order of the inductor L_1, $f(V_1)$ is the piecewise-linear $v-i$ characteristic of nonlinear Chua's diode, which can be described by (5.2).

By using the transformation (5.3), we can rewrite Eqs. (5.12) into the following dimensionless form (Petráš, 2008):

$$
\begin{aligned}
{}_0D_t^{q_1} x(t) &= \alpha\left(y(t) - x(t) - f(x)\right), \\
{}_0D_t^{q_2} y(t) &= x(t) - y(t) + z(t), \\
{}_0D_t^{q_3} z(t) &= -\beta y(t) - \gamma z(t),
\end{aligned}
\tag{5.13}
$$

where $f(x)$ is the piecewise-linear nonlinearity (5.5).

5.2.2.1 Experimental Measurements

Classical Chua's oscillator can also be realized by electrical elements according to the scheme shown in Fig. 5.3, which is a very simple electronic circuit that exhibits nonlinear dynamical phenomena such as bifurcation and chaos (Kennedy, 1992). Chua's diode (5.2) – nonlinear resistor – was realized by operating amplifier LM 358 and resistors R_1, R_7, and R_8 ($R_7 = R_8$) as negative impedance converter (Bartissol and Chua, 1988).

For experimental verification of Chua's system depicted in Fig. 5.3 and described by Eqs. (5.12) and (5.2), the following values of electrical elements were chosen (Caponetto et al., 2010; Petráš, 2008):

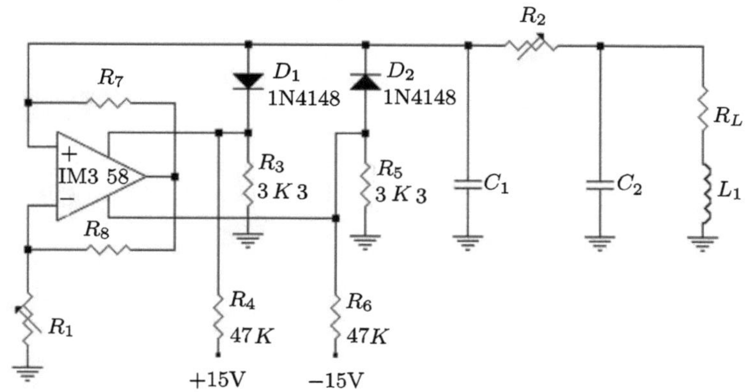

Fig. 5.3 Practical realization of Chua's circuit.

$$C_1 = 4.71\,nF,\ C_2 = 48\,nF,\ L_1 = 4.64\,mH,$$
$$R_L = 15.8\,\Omega,\ R_1 = 897\,\Omega,\ R_2 = 998\,\Omega,\ R_7 = R_8 = 393\,\Omega. \tag{5.14}$$

We use the metalized paper capacitors C_1 and C_2 with the real order $q_1 = q_2 = 0.98$ and we assume the real order of inductor $q_3 = 0.94$ (see e.g. (Schafer and Kruger, 2008; Westerlund and Ekstam, 1994; Westerlund, 2002)). The total order of the system is $\bar{q} = 2.90$.

The measured breakpoints of the non-linear characteristic (5.2) are:

$$-B_p = (-8.79\,V,\ 7.7\,mA),\quad B_p = (9.12\,V,\ -7.9\,mA).$$

Assuming the three-segment piecewise-linear voltage-current transfer characteristic of negative impedance converter (5.2), we have the slope $G_a = -1/R_1 = -1.1148\ mA/V$ for $R_7 = R_8$ and the slope G_b was calculated using the breakpoints B_p and it has the value $G_b = -0.8710\,mA/V$.

The resistors R_3, R_4, R_5, R_6, and the diodes D_1 and D_2 generate the positive and negative halves of the nonlinearity.

In Fig. 5.4 is depicted the photo of the digital oscilloscope screen (Tektronix TDS1002, 60 Mhz). It is a real measurement of voltages $V_1 - V_2$ for circuit presented in Fig. 5.3 with the parameters of electrical components (5.14). The result shown in Fig. 5.4 is the double-scroll attractor of fractional-order Chua's system described by Eqs. (5.12) and (5.2). We can observe an amplification of the system.

An alternative scheme of the practical implementation of the Chua's oscillator with two operating amplifiers for a different kind of nonlinearity with saturation can be found, for instance, in (Chua et al., 1993; Kennedy, 1993b) or for an IC chip implementation we refer to (Cruz, 1993).

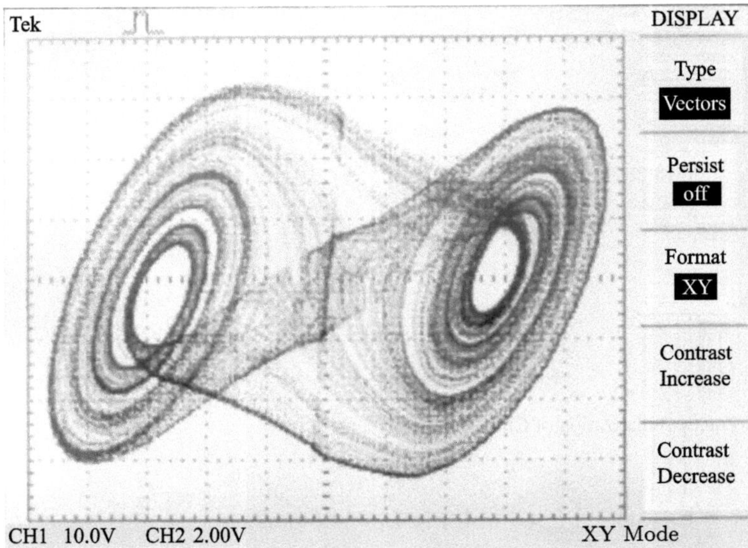

Fig. 5.4 Photo of oscilloscope screen: Strange attractor of the Chua's system (5.12).

5.2.2.2 Simulation Results

For simulation purposes we will use a numerical solution of Chua's equations (5.13) obtained by using the relationship (2.53) derived from the Grünwald-Letnikov definition (2.15), which leads to equations in the form:

$$\begin{aligned}
x(t_k) &= (\alpha(y(t_{k-1}) - x(t_{k-1}) - f(x(t_{k-1}))))\,h^{q_1} - \sum_{j=v}^{k} c_j^{(q_1)} x(t_{k-j}),\\
y(t_k) &= (x(t_k) - y(t_{k-1}) + z(t_{k-1}))\,h^{q_2} - \sum_{j=v}^{k} c_j^{(q_2)} y(t_{k-j}),\\
z(t_k) &= (-\beta y(t_k) - \gamma z(t_{k-1}))\,h^{q_3} - \sum_{j=v}^{k} c_j^{(q_3)} z(t_{k-j}),
\end{aligned} \tag{5.15}$$

where

$$f(x(t_{k-1})) = m_1 x(t_{k-1}) + \frac{1}{2}(m_0 - m_1) \times (|x(t_{k-1}) + 1| - |x(t_{k-1}) - 1|) \tag{5.16}$$

and where T_{sim} is the simulation time, $k = 1, 2, 3, \ldots, N$, for $N = [T_{sim}/h]$, and $(x(0)$, $y(0)$, $z(0))$ is the start point (initial conditions). The binomial coefficients $c_j^{(q_i)}$, $\forall i$, are calculated according to relation (2.54).

Similar and comparable results we have measured can be obtained by simulation using Eqs. (5.15) for time step $h = 0.001$ and the short memory principle with length $L_m = 10$ (10000 values and coefficients from history). Figure 5.5 shows the double-

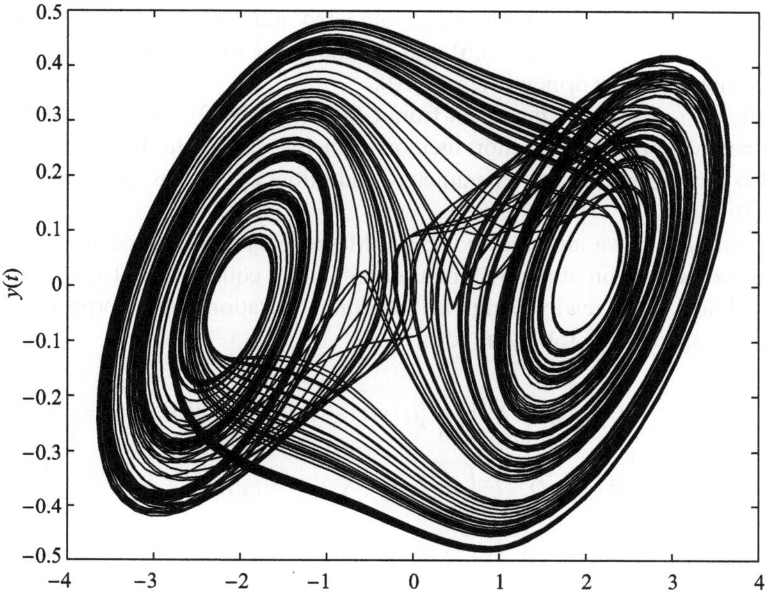

Fig. 5.5 Strange attractor of the fractional-order Chua's system (5.13) with total order $\bar{q} = 2.90$ for the parameters: $\alpha = 10.1911$, $\beta = 10.3035$, $\gamma = 0.1631$, $q_1 = q_2 = 0.98$, $q_3 = 0.94$, $m_0 = -1.1126$, $m_1 = -0.8692$, and simulation time $T_{sim} = 100\,s$.

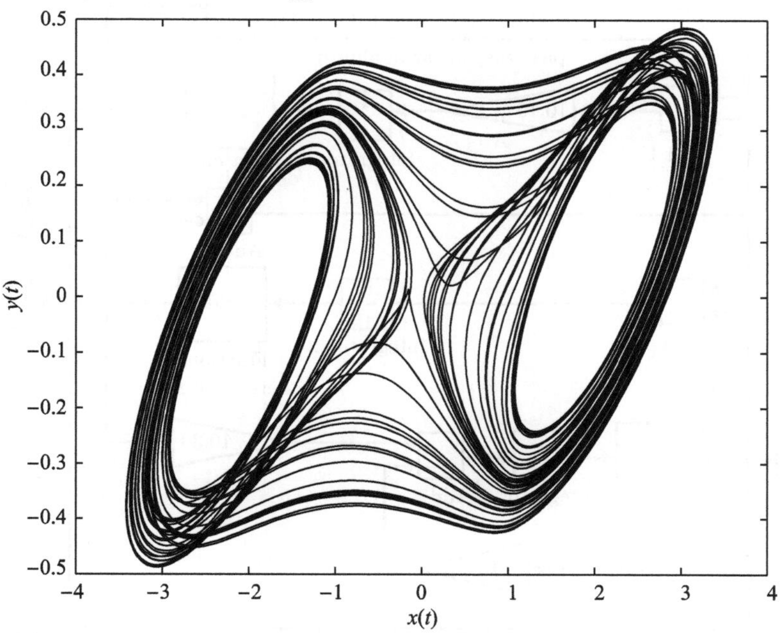

Fig. 5.6 Strange attractor from the fractional-order Chua's system (5.13) with total order $\bar{q} = 2.84$ for the parameters: $\alpha = 10.725$, $\beta = 10.593$, $\gamma = 0.268$, $q_1 = 0.93$, $q_2 = 0.99$, $q_3 = 0.92$, $m_0 = -1.1726$, $m_1 = -0.7872$ and simulation time $T_{sim} = 100\,s$.

scroll attractor of Chua's circuit (5.13) computed numerically for initial conditions $(x(0), y(0), z(0)) = (0.6, 0.1, -0.6)$ and for the value of electrical parts (5.14) by using a short memory principle ($L_m = 10$).

In Fig. 5.6 is depicted the result from (Zhu et al., 2009), where simulation was performed without using the short memory principle ($\nu = 1$) for time step $h = 0.001$ and also for a different set of parameters and initial conditions $(x(0), y(0), z(0)) = (0.2, -0.1, 0.1)$.

For simulation we are able to use the Matlab/Simulink approach as well. The state-space expression of the fractional-order Chua's equations (5.13) with parameters α, β, and γ is given by using the integration operation and the properties (2.50) and (2.51), and has the following form:

$$
\begin{aligned}
x(t) &= {}_0D_t^{1-q_1}\left(\int_0^t [\alpha(y(t) - x(t) - f(x))]\,dt\right), \\
y(t) &= {}_0D_t^{1-q_2}\left(\int_0^t [x(t) - y(t) + z(t)]\,dt\right), \\
z(t) &= {}_0D_t^{1-q_3}\left(\int_0^t [-\beta y(t) - \gamma z(t)]\,dt\right).
\end{aligned}
\tag{5.17}
$$

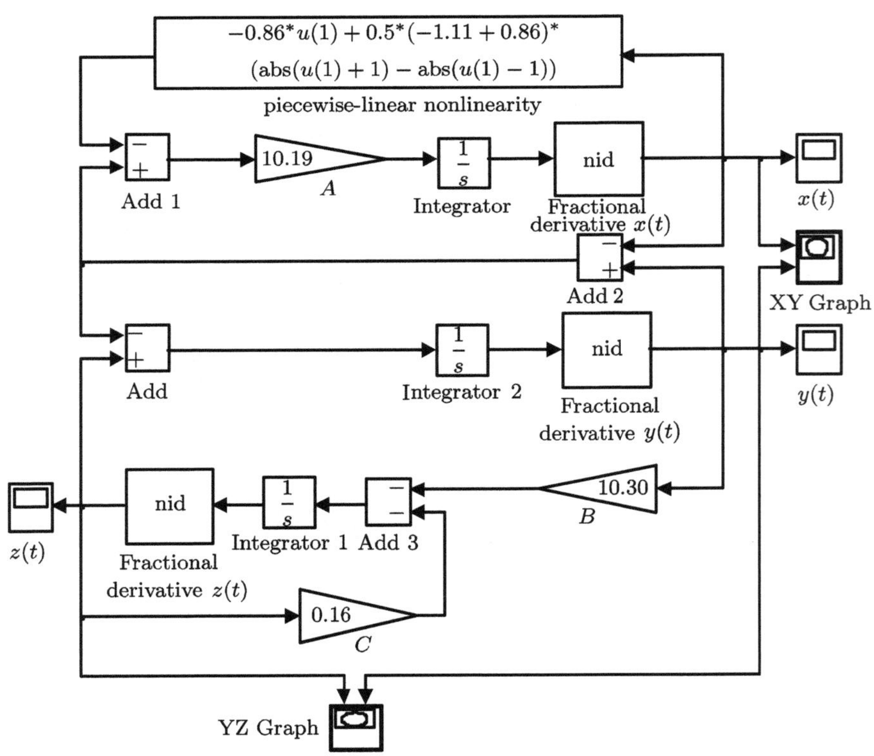

Fig. 5.7 Matlab/Simulink block diagram (model) for Chua's system (5.13).

The system model developed from the state equations (5.17) for system parameters α, β, and γ by using the Matlab/Simulink environment is depicted in Fig. 5.7. For simulation of the fractional derivative (integral) we used a Simulink block *nid* created by Duarte Valerio (Valerio, 2005) with the combination of the classical integrator by using the property of commutation of two operators.

In Fig. 5.8 is depicted the simulation results obtained by numerical simulation in the Matlab/Simulink for the following values of the parameters: $A \equiv \alpha = 10.19$, $B \equiv \beta = 10.30$, $C \equiv \gamma = 0.16$, $q_1 = q_2 = 0.98$, and $q_3 = 0.94$, for the initial conditions: $(x(0), y(0), z(0)) = (0.6, 0.1, -0.6)$ and for the slopes of Chua's diode (5.5) characteristic: $m_0 = -1.11$ and $m_1 = -0.86$.

As we can observe in Fig. 5.8, the obtained simulation results are comparable to the results depicted in Fig. 5.5.

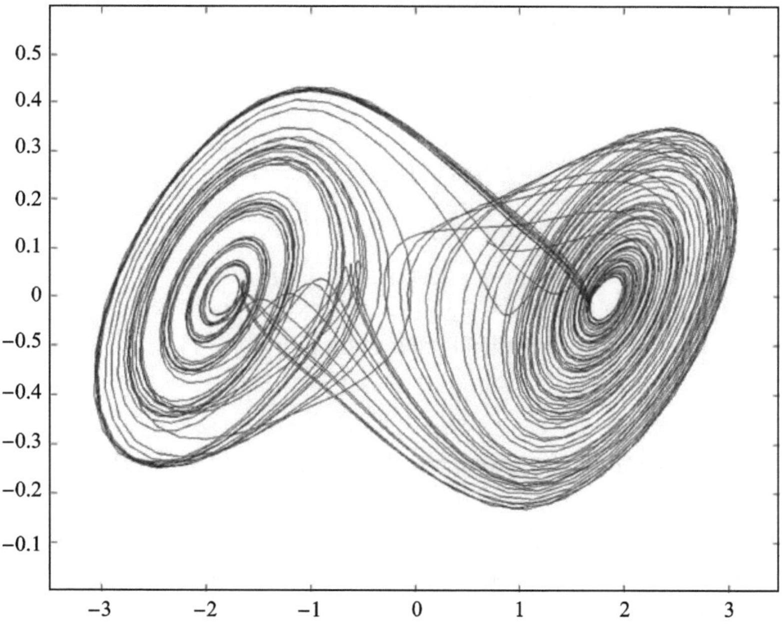

Fig. 5.8 Simulation result (x vs. y) of Chua's system (5.13).

5.2.3 Fractional-Order Chua-Podlubny's Oscillator

This system uses an approach where the order of any of three constitutive equations (5.4) can be changed so that the total order gives the desired value. In Chua-Podlubny's case, in equation one the first differentiation is replaced by a fractional one. The final dimensionless equations of the system are (Podlubny, 1999):

$$
\begin{aligned}
{}_0D_t^q x(t) &= \alpha\, {}_0D_t^{q-1}\left(y(t)-x(t)\right) - \frac{2\alpha}{7}\left(4x(t)-x^3(t)\right), \\
\frac{dy(t)}{dt} &= x(t)-y(t)+z(t), \\
\frac{dz(t)}{dt} &= -\frac{100}{7}y(t) = -\beta y(t),
\end{aligned}
\tag{5.18}
$$

where $\alpha = C_2/C_1$ and $\beta = C_2R_2^2/L_1$, and fractional order $q < 1$, $q \in \mathrm{R}$.

5.2.4 Fractional-Order Chua-Hartley's Oscillator

The Chua-Hartley's system is different from the usual Chua's system (5.4) in that the piecewise-linear nonlinearity is replaced by an appropriate cubic nonlinearity which yields very similar behavior. Derivatives on the left side of the differential equations are replaced by the fractional derivatives as follows (Hartley et al., 1995):

$$
\begin{aligned}
{}_0D_t^q x(t) &= \alpha\left(y(t) + \frac{x(t)-2x^3(t)}{7}\right), \\
{}_0D_t^q y(t) &= x(t)-y(t)+z(t), \\
{}_0D_t^q z(t) &= -\beta y(t) = -\frac{100}{7}y(t),
\end{aligned}
\tag{5.19}
$$

where $q \leq 1$, $q \in \mathrm{R}$ is the fractional order of derivatives.

5.2.5 Fractional-Order Memristor-Based Chua's Oscillator

Since the memristor was postulated by L. O. Chua in 1971 and discovered by R. Williams et al. (HP laboratory) in 2008 (realized as a $Pt - TiO_2 - Pt$ device), it becomes the fourth circuit element. This fact allows us to use the memristor as a nonlinear element in circuits which exhibit chaos. In the case of Chua's circuit, the nonlinear resistor NR is replaced by an active memristor M as shown in Fig. 5.9.

The memristor in Fig. 5.9 is a flux-controlled memristor whose characteristic is given by literature (Chua, 1971):

$$I_M(t) = W(\phi(t))V_1(t), \tag{5.20}$$

where $W(\phi(t))$ is the incremental memductance defined as (Chua, 1971)

$$W(\phi) = \frac{dq(\phi)}{d\phi}. \tag{5.21}$$

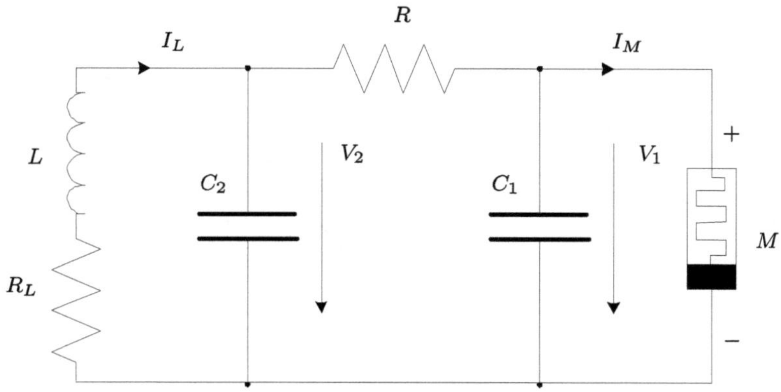

Fig. 5.9 Chua's circuit with active memristor.

Taking this into account we can write the equations for the memristor-based chaotic circuit depicted in Fig. 5.9 as follows:

$$\begin{aligned} \frac{dV_1(t)}{dt} &= \frac{1}{C_1}\left[G(V_2(t)-V_1(t))-I_M(t))\right], \\ \frac{dV_2(t)}{dt} &= \frac{1}{C_2}\left[G(V_1(t)-V_2(t))+I_L(t)\right], \\ \frac{dI_L(t)}{dt} &= \frac{1}{L}\left[-V_2(t)-R_L I_L(t)\right], \\ \frac{d\phi(t)}{dt} &= V_1(t), \end{aligned} \tag{5.22}$$

where $G = 1/R$, and $I_M(t)$ is defined by Eq. (5.20).

For the flux-controlled memristor a monotone-increasing piecewise-linear characteristic (Itoh and Chua, 2008, 2009) was assumed. The memristor constitutive relation is shown in Fig. 5.10 and can be expressed as

$$q(\phi) = b\phi + 0.5(a-b) \times (|\phi+1| - |\phi-1|), \tag{5.23}$$

where $a, b > 0$. The memductance function obtained from the $q(\phi)$ function is:

$$W(\phi) = \frac{dq(\phi)}{d\phi} \begin{cases} a, & |\phi| < 1, \\ b, & |\phi| > 1. \end{cases} \tag{5.24}$$

Several modifications of the memristor-based Chua's circuit, where chaos was observed, were described and analyzed in (Itoh and Chua, 2008). For a practical implementation of the memristor it is possible to use operating amplifiers (Zhong, 1994) and then a smooth cubic nonlinearity could be used for replacing a $q-\phi$ function depicted in Fig. 5.10.

The dynamics of the Chua's circuit with a passive memristor (flux-controlled memristor and negative conductance) depicted in Fig. 5.11 are given by the following set of differential equations:

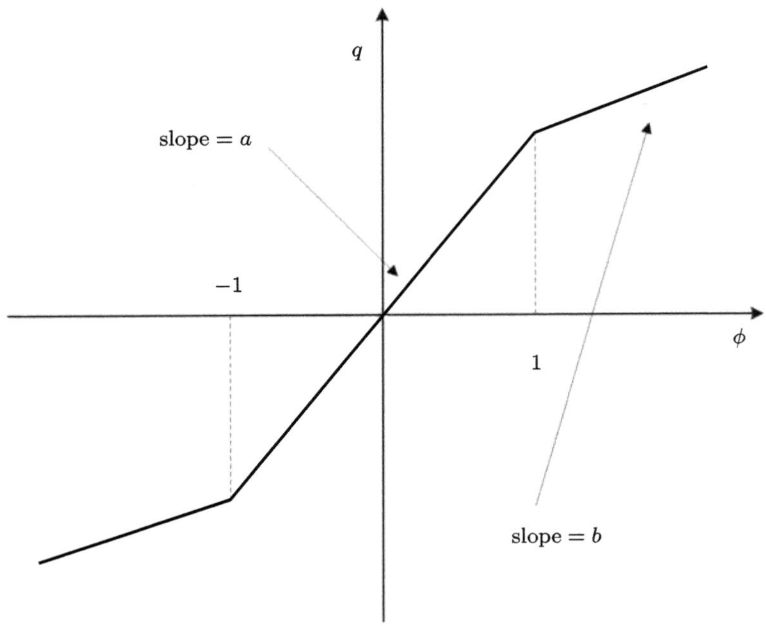

Fig. 5.10 The constitutive relation of a piecewise-linear flux-controlled memristor.

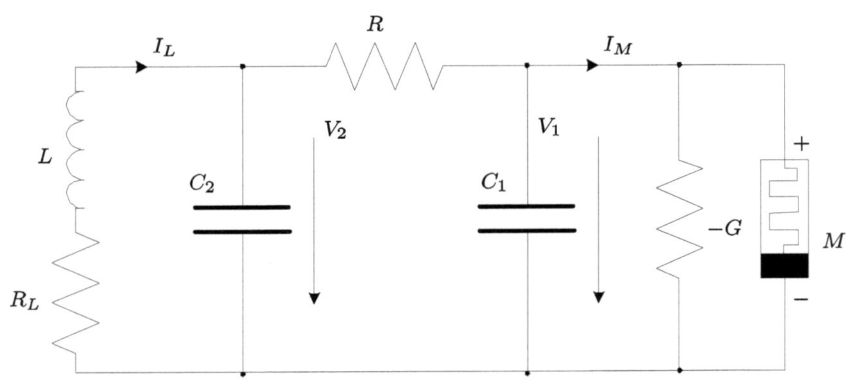

Fig. 5.11 Chua's circuit with flux-controlled memristor and negative conductance.

$$
\begin{aligned}
\frac{dV_1(t)}{dt} &= \frac{1}{C_1}\left[\frac{(V_2(t)-V_1(t))}{R} + GV_1(t) - W(\phi)V_1(t))\right],\\
\frac{dV_2(t)}{dt} &= \frac{1}{C_2}\left[\frac{(V_1(t)-V_2(t))}{R} + I_L(t)\right],\\
\frac{dI_L(t)}{dt} &= \frac{1}{L}\left[-V_2(t) - R_L I_L(t)\right],\\
\frac{d\phi(t)}{dt} &= V_1(t),
\end{aligned}
\tag{5.25}
$$

where functions $q(\phi)$ and $W(\phi)$ are given in (5.23) and (5.24), respectively.

If we set

$$\begin{aligned} &x = V_1, \quad y = V_2, \quad z = I_L, \quad w = \phi, \quad C_2 = 1, \quad R = 1, \\ &\alpha = 1/C_1, \quad \beta = 1/L, \quad \gamma = R_L/L, \quad \zeta = G, \end{aligned} \tag{5.26}$$

then Eqs. (5.25) can be transformed into the dimensionless form (Itoh and Chua, 2008):

$$\begin{aligned} \frac{dx(t)}{dt} &= \alpha\left(y(t) - x(t) + \zeta x(t) - W(w)x(t)\right), \\ \frac{dy(t)}{dt} &= x(t) - y(t) + z(t), \\ \frac{dz(t)}{dt} &= -\beta y(t) - \gamma z(t), \\ \frac{dw(t)}{dt} &= x(t), \end{aligned} \tag{5.27}$$

where piecewise-linear function $W(w)$ is given below:

$$W(w) = \begin{cases} a: & |w| < 1, \\ b: & |w| > 1. \end{cases} \tag{5.28}$$

The equilibrium points of the system (5.27) are given by setting the left side of equations to 0 except the last one. We set w as constant, which corresponds to the w–axis (Itoh and Chua, 2008). The Jacobian matrix at this equilibrium state is (Itoh and Chua, 2008):

$$\mathbf{J}_W = \begin{bmatrix} \alpha(-1+\zeta-W(w)) & \alpha & 0 & 0 \\ 1 & -1 & 1 & 0 \\ 0 & -\beta & -\gamma & 0 \\ 1 & 0 & 0 & 0 \end{bmatrix}. \tag{5.29}$$

For the following parameter set (Itoh and Chua, 2008): $\alpha = 10, \;\; \beta = 13, \;\; \gamma = 0.35, \;\; \zeta = 1.5, \;\; a = 0.3$, and $b = 0.8$, four eigenvalues $\lambda_i \;\; (i = 1,2,3,4)$ for $|w| < 1$ can be written as

$$\lambda_{1,2} \approx -1.31103 \pm 2.74058j, \; \lambda_3 \approx 3.27207, \; \lambda_4 = 0$$

and four eigenvalues for $|w| > 1$ can be written as

$$\lambda_{1,2} \approx 0.07865 \pm 2.84655j, \; \lambda_3 \approx -4.50731, \; \lambda_4 = 0.$$

They are characterized by an unstable saddle-focus point and numerical simulations for the above paremeters show that the system (5.27) has chaotic behavior. In Fig. 5.12 and Fig. 5.13 are depicted chaotic attractors in 3D state space for $T_{sim} = 200\,s$.

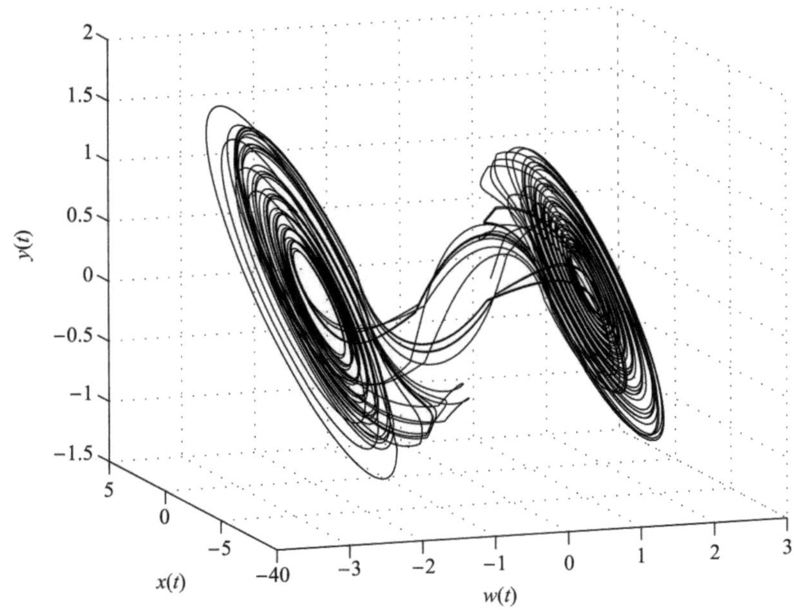

Fig. 5.12 Strange attractor of the memristor-based Chua's system (5.27) in $w-x-y$ state space, for parameters $\alpha = 10$, $\beta = 13$, $\gamma = 0.35$, $\zeta = 1.5$, $a = 0.3$, and $b = 0.8$, initial conditions: $x(0) = 0.8, y(0) = 0.05, z(0) = 0.007, w(0) = 0.6$ and simulation time $T_{sim} = 200\,s$.

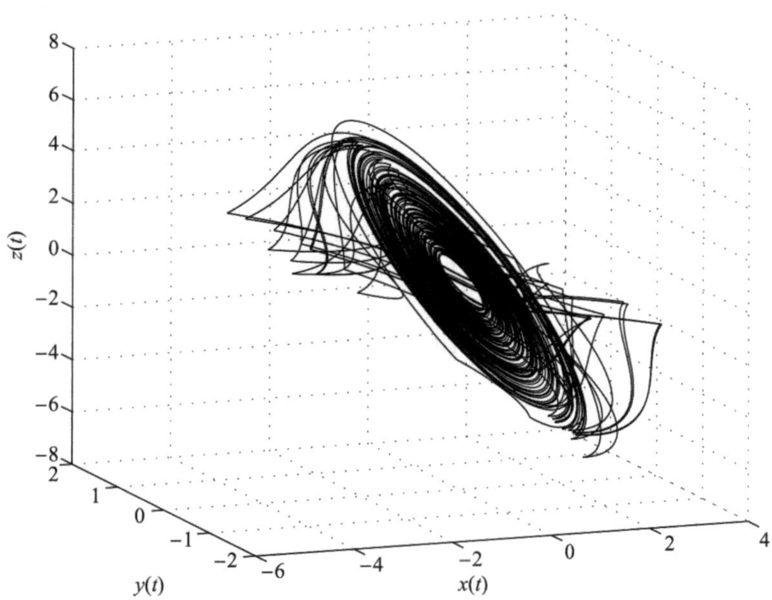

Fig. 5.13 Strange attractor of the memristor-based Chua's system (5.27) in $x-y-z$ state space, for parameters $\alpha = 10$, $\beta = 13$, $\gamma = 0.35$, $\zeta = 1.5$, $a = 0.3$, and $b = 0.8$, initial conditions: $x(0) = 0.8, y(0) = 0.05, z(0) = 0.007, w(0) = 0.6$ and simulation time $T_{sim} = 200\,s$.

In Fig.5.14 and Fig. 5.15 are depicted the attractors of the memristor-based Chua's system (5.27) for parameters $\alpha = 10$, $\beta = 13$, $\gamma = 0.35$, $\zeta = 1.5$, $a = 0.3$, $b = 0.8$, initial conditions: $x(0) = 0.8, y(0) = 0.05, z(0) = 0.007, w(0) = 0.6$ and simulation time $T_{sim} = 200\,s$, projected onto $y - w$, and $z - w$ planes, respectively.

If we consider a fractional-order model for each electrical element in the circuit depicted in Fig. 5.11, we can write a more general mathematical model for this circuit. As already mentioned, real capacitor and real inductor are "fractional" and for real memristor we postulated a fractional-order model as well $(d^{\alpha}\phi(t)/dt^{\alpha} = V(t))$. By using a technique of fractional calculus we obtain the following equations:

$$\begin{aligned}
{}_0D_t^{q_1}x(t) &= \alpha\left(y(t) - x(t) + \zeta x(t) - W(w)x(t)\right),\\
{}_0D_t^{q_2}y(t) &= x(t) - y(t) + z(t),\\
{}_0D_t^{q_3}z(t) &= -\beta y(t) - \gamma z(t),\\
{}_0D_t^{q_4}w(t) &= x(t),
\end{aligned} \tag{5.30}$$

where function $W(w)$ is given in (5.28) and where q_1, q_2, q_3, and q_4 are fractional orders of real electrical elements (memristive systems): capacitor C_1, capacitor C_2, inductor L, and memristor M, respectively.

The stability of the new fractional-order memristor-based Chua's system can be investigated by using Theorem 4.6. For the fractional incommensurate-order system (5.30), we can rewrite real order as $q_i = v_i/u_i$, $v_i, u_i \in Z^+$ for $i = 1,2,3,4$ and if we set $\gamma = 1/m$, where m is the LCM of the denominators, the characteristic equation of the system (5.30) for the Jacobian matrix $\mathbf{J}_W$ is:

$$\det(\mathrm{diag}([\lambda^{mq_1}\,\lambda^{mq_2}\,\lambda^{mq_3}\,\lambda^{mq_4}]) - \mathbf{J}_W) = 0$$

and then the stability condition is defined as follows:

$$|\arg(\lambda_i)| > \gamma\frac{\pi}{2}$$

for all eigenvalues λ_i.

In the case of piecewise-nonlinearity depicted in Fig. 5.10, we should investigate the characteristic equation for the linear part with slope a and for the linear part with slope b. If $|w| < 1$ then we are dealing with slope a and the Jacobian matrix is

$$\mathbf{J}_{Wa} = \begin{bmatrix} \alpha(-1+\zeta-a) & \alpha & 0 & 0\\ 1 & -1 & 1 & 0\\ 0 & -\beta & -\gamma & 0\\ 1 & 0 & 0 & 0 \end{bmatrix} \tag{5.31}$$

and if $|w| > 1$ then we are dealing with slope b and the Jacobian matrix is

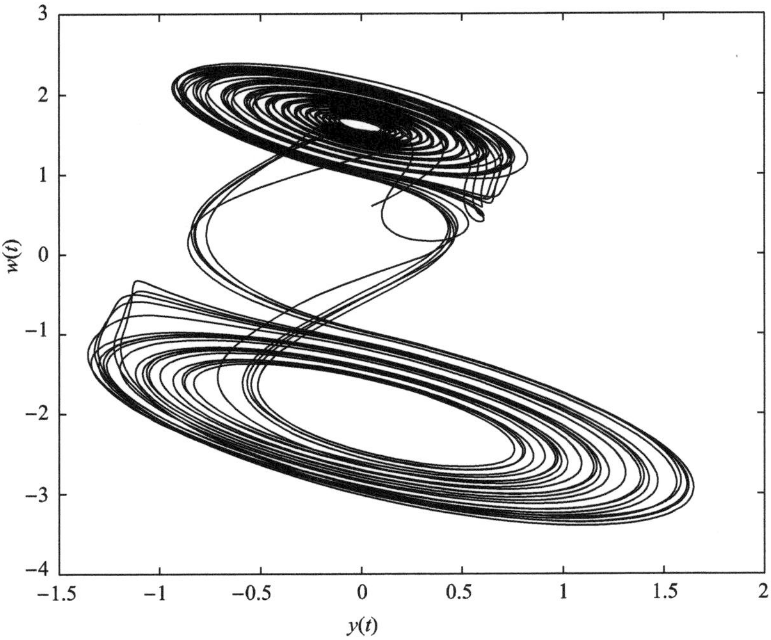

Fig. 5.14 Strange attractor of the memristor-based Chua's system projected onto $y - w$ plane.

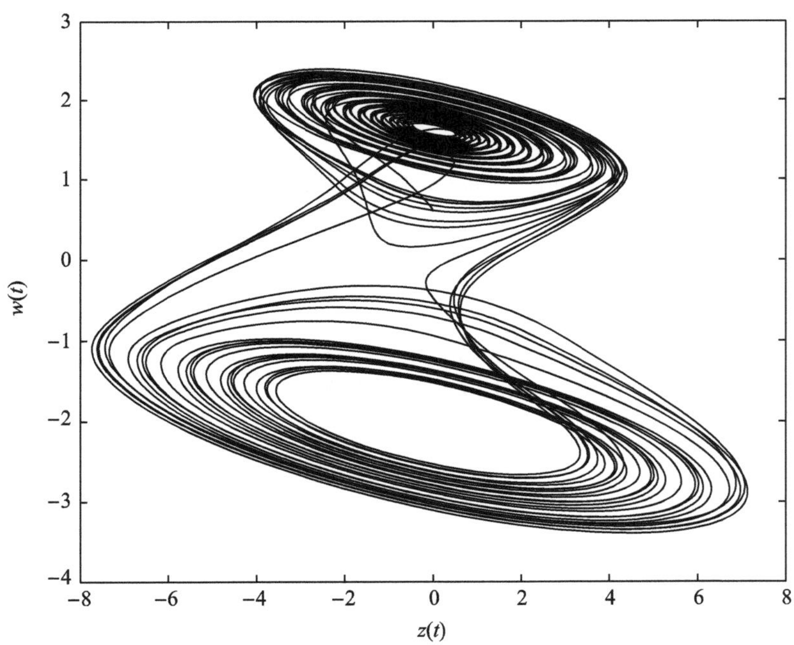

Fig. 5.15 Strange attractor of the memristor-based Chua's system projected onto $z - w$ plane.

$$\mathbf{J}_{Wb} = \begin{bmatrix} \alpha(-1+\zeta-b) & \alpha & 0 & 0 \\ 1 & -1 & 1 & 0 \\ 0 & -\beta & -\gamma & 0 \\ 1 & 0 & 0 & 0 \end{bmatrix}. \tag{5.32}$$

The characteristic equation for the linear part with the slope a with matrix (5.31) is

$$\det(\mathrm{diag}([\lambda^{mq_1}\,\lambda^{mq_2}\,\lambda^{mq_3}\,\lambda^{mq_4}]) - \mathbf{J}_{Wa}) = 0$$

and for linear part with the slope b and matrix (5.32) it has the form

$$\det(\mathrm{diag}([\lambda^{mq_1}\,\lambda^{mq_2}\,\lambda^{mq_3}\,\lambda^{mq_4}]) - \mathbf{J}_{Wb}) = 0.$$

When we consider a simple case where the fractional-order memristor-based Chua's system has commensurate order, which means $q_1 = q_2 = q_3 = q_4 \equiv q$, the stability can be investigated according to Theorem 4.5, where the condition is:

$$|\arg(\mathrm{eig}(\mathbf{J}_W))| = |\arg(\lambda_i)| > q\frac{\pi}{2}$$

for all eigenvalues λ_i.

As in the previous case, stability should be investigated for both piecewise-linear parts of memristor characteristic shown in Fig. 5.10. In this case it means that we should find the angle of all eigenvalues for both Jacobian matrices $\mathbf{J}_{Wa}$ and $\mathbf{J}_{Wb}$ respectively. A necessary stability condition for fractional-order systems (5.30) to remain chaotic is keeping at least one eigenvalue λ in the unstable region.

Because the frequency approximation techniques (Carlson and Halijak, 1963; Oustaloup et al., 2000) are unreliable in recognising chaos in fractional-order nonlinear systems (Tavazoei and Haeri, 2007a), for simulation purposes we use a numerical solution of the memristor-based Chua's equations (5.30) obtained by the method descibed in (Petráš, 2009a, 2010). It is a time domain method derived by using the relationship (2.53), which leads to equations in the following form:

$$\begin{aligned} x(t_k) &= (\alpha(y(t_{k-1}) - x(t_{k-1}) + \zeta x(t_{k-1}) - W(w(t_{k-1}))x(t_{k-1})))\,h^{q_1} - \\ &\quad - \sum_{j=v}^{k} c_j^{(q_1)} x(t_{k-j}), \\ y(t_k) &= (x(t_k) - y(t_{k-1}) + z(t_{k-1}))\,h^{q_2} - \sum_{j=v}^{k} c_j^{(q_2)} y(t_{k-j}), \\ z(t_k) &= (-\beta y(t_k) - \gamma z(t_{k-1}))\,h^{q_3} - \sum_{j=v}^{k} c_j^{(q_3)} z(t_{k-j}), \\ w(t_k) &= x(t_k) h^{q_4} - \sum_{j=v}^{k} c_j^{(q_4)} w(t_{k-j}), \end{aligned} \tag{5.33}$$

where

$$W(w(t_{k-1})) = a \quad \text{for} \quad |w(t_{k-1})| < 1, \\ W(w(t_{k-1})) = b \quad \text{for} \quad |w(t_{k-1})| > 1, \tag{5.34}$$

and where T_{sim} is the simulation time, $k = 1, 2, 3, \ldots, N$, for $N = [T_{sim}/h]$, and ($x(0)$, $y(0)$, $z(0)$, $w(0)$) is the start point (initial conditions).

The binomial coefficients $c_j^{(q_i)}$, $\forall i$ are calculated according to relation (2.54).

If we consider the parameter set $\alpha = 10$, $\beta = 13$, $\gamma = 0.35$, $\zeta = 1.5$, $a = 0.3$, and $b = 0.8$, according to Definition 4.5, for these parameters we are able to calculate a minimal commensurate order for which the system (5.30) remains chaotic. In this case it is $q > 0.98$.

Let us consider the paremeter set $\alpha = 10$, $\beta = 13$, $\gamma = 0.1$, $\zeta = 1.5$, $a = 0.3$, and $b = 0.8$. For these parameters a minimal commensurate order is $q > 0.95$. We performed a simulation for the above parameters and commensurate order $q = 0.97$ ($q_1 = q_2 = q_3 = q_4 = 0.97$). The total order of the system is 3.88.

In Fig. 5.16 and Fig. 5.17 are depicted chaotic attractors in 3D state space for $T_{sim} = 200\,s$. Both simulations were performed without using the short memory principle ($v = 1$) for time step $h = 0.005$ with the initial conditions: $x(0) = 0.8$, $y(0) = 0.05$, $z(0) = 0.007$, $w(0) = 0.6$.

In Fig.5.18 and Fig. 5.19 are depicted the attractors of the memristor-based Chua's system (5.30) for parameters (5.35), $a = 0.3$, $b = 0.8$, orders $q_1 = q_2 = q_3 = q_4 = 0.97$, initial conditions: $x(0) = 0.8$, $y(0) = 0.05$, $z(0) = 0.007$, $w(0) = 0.6$ and simulation time $T_{sim} = 100\,s$, projected onto $y - w$, and $z - w$ planes, respectively.

When we consider real orders of capacitors models (Westerlund and Ekstam, 1994): $q_1 = q_2 = 0.98$, real order of inductor model (Schafer and Kruger, 2008): $q_3 = 0.99$, and we assume a real order of memristor model: $q_4 = 0.97$, for the parameters:

$$\alpha = 10, \quad \beta = 13, \quad \gamma = 0.1, \quad \zeta = 1.5, \tag{5.35}$$

$a = 0.3$, $b = 0.8$, the initial conditions: $x(0) = 0.8$, $y(0) = 0.05$, $z(0) = 0.007$, $w(0) = 0.6$, simulation time $T_{sim} = 100\,s$, and time step $h = 0.005$, we get the chaotic double-scroll attractor as well for the total system order 3.92.

In Fig. 5.20 and Fig. 5.21 are depicted chaotic attractors in 3D state space for $T_{sim} = 100\,s$. The simulations were performed without using the short memory principle ($v = 1$) for time step $h = 0.005$ with the initial conditions: $x(0) = 0.8$, $y(0) = 0.05$, $z(0) = 0.007$, $w(0) = 0.6$. In this case, we just estimated the real order of the memristor. Simulations show double-scroll atractors and we can observe a chaotic behavior.

The characteristic equation of the system (5.30) with parameters (5.35), orders $q_1 = q_2 = 0.98 = 98/100$, $q_3 = 0.99 = 99/100$, $q_4 = 0.97 = 97/100$, for Jacobian (5.31) is

$$\lambda^{392} - \lambda^{294} + \frac{\lambda^{293}}{10} - 12\lambda^{196} + \frac{129\lambda^{195}}{10} - \frac{136\lambda^{97}}{5} = 0$$

and for Jacobian (5.32) it has form

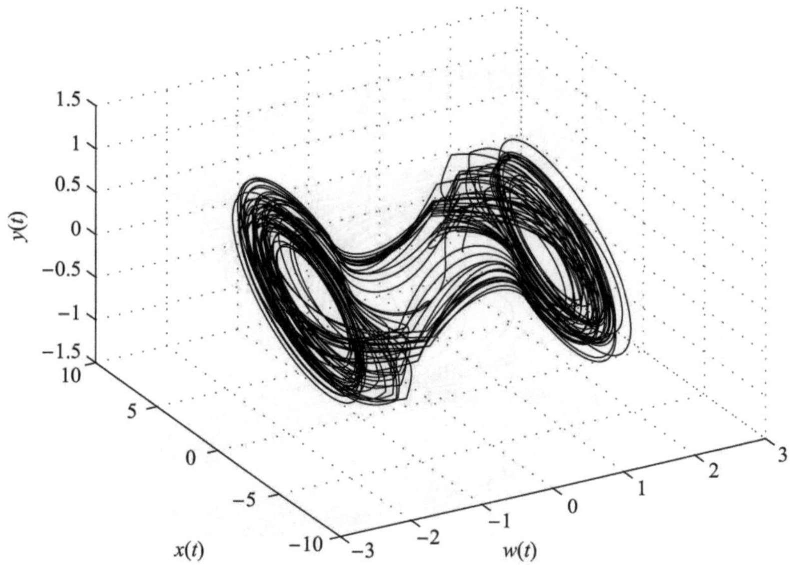

Fig. 5.16 Strange attractor of the memristor-based Chua's system (5.30) in $w-x-y$ state space, for parameters $\alpha = 10$, $\beta = 13$, $\gamma = 0.1$, $\zeta = 1.5$, $a = 0.3$, $b = 0.8$, orders $q_1 = q_2 = q_3 = q_4 = 0.97$, initial conditions: $x(0) = 0.8$, $y(0) = 0.05$, $z(0) = 0.007$, $w(0) = 0.6$ and simulation time $T_{sim} = 200\,s$.

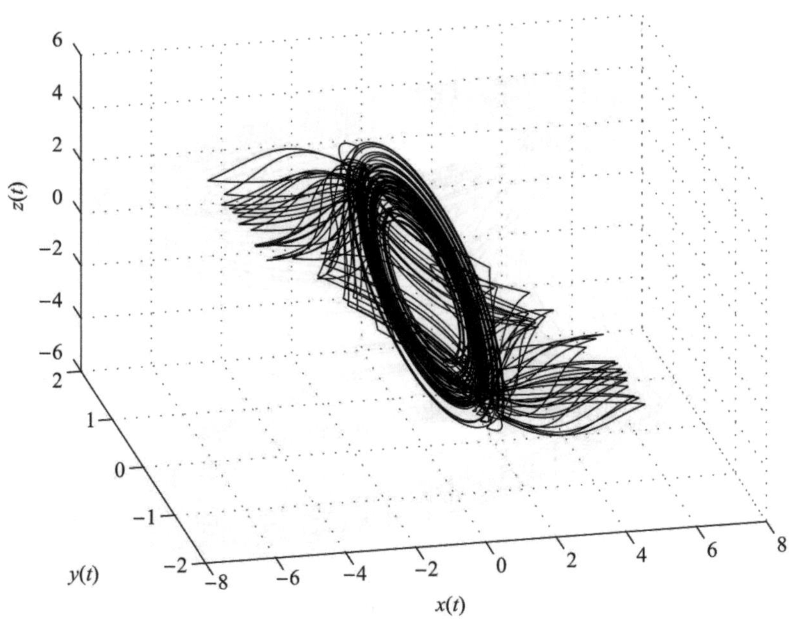

Fig. 5.17 Strange attractor of the memristor-based Chua's system (5.30) in $x-y-z$ state space, for parameters $\alpha = 10$, $\beta = 13$, $\gamma = 0.1$, $\zeta = 1.5$, $a = 0.3$, $b = 0.8$, orders $q_1 = q_2 = q_3 = q_4 = 0.97$, initial conditions: $x(0) = 0.8$, $y(0) = 0.05$, $z(0) = 0.007$, $w(0) = 0.6$ and simulation time $T_{sim} = 200\,s$.

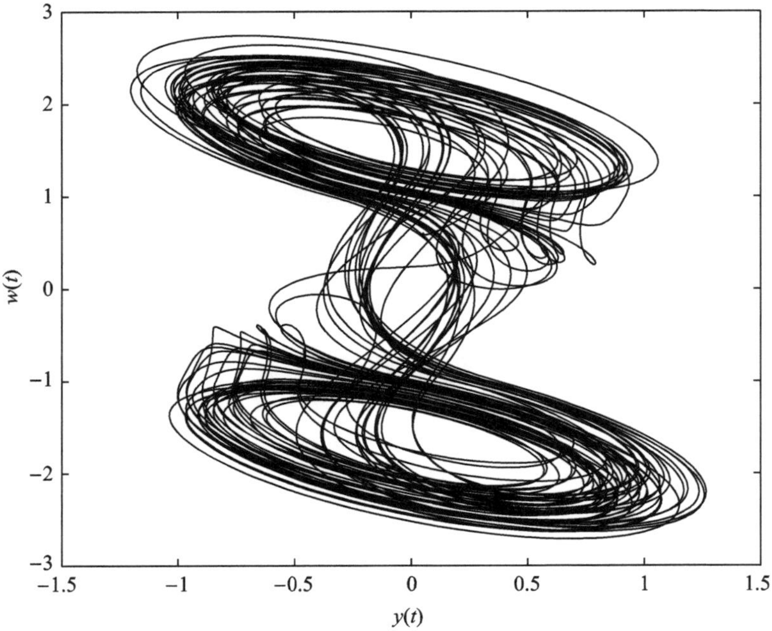

Fig. 5.18 Strange attractor of the memristor-based Chua's system projected onto $y-w$ plane.

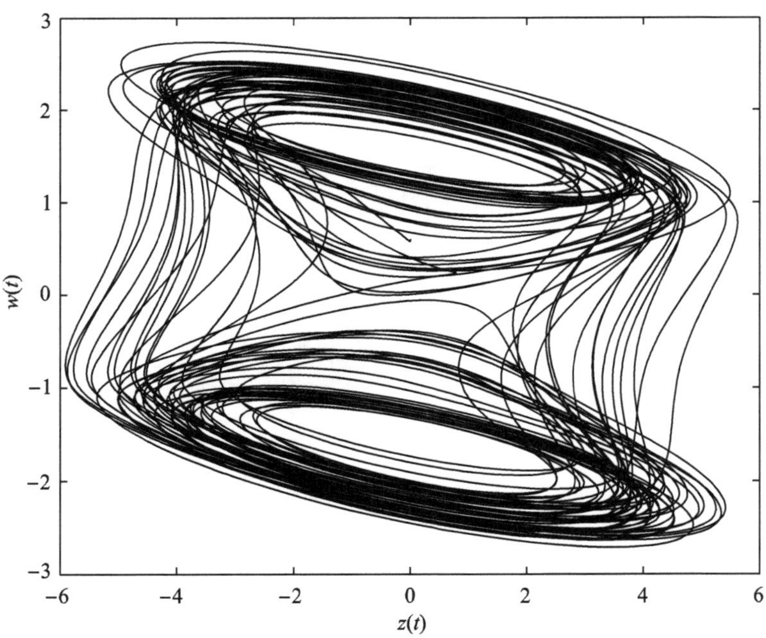

Fig. 5.19 Strange attractor of the memristor-based Chua's system projected onto $z-w$ plane.

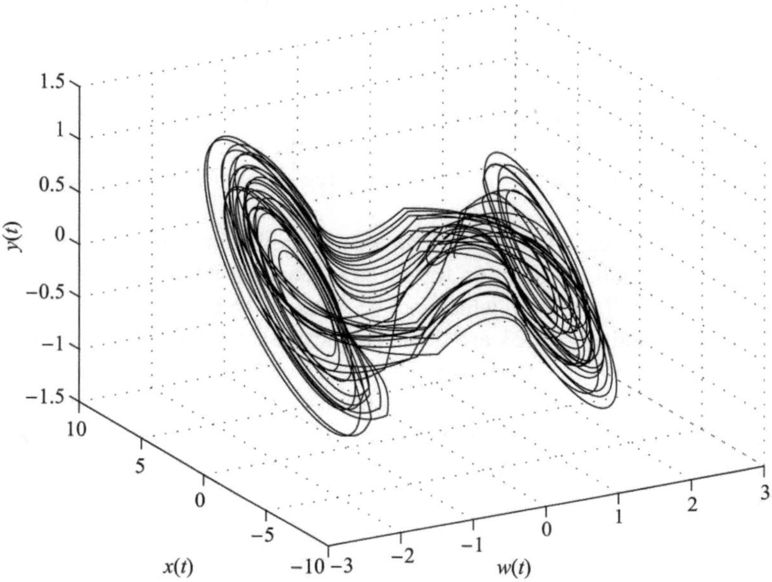

Fig. 5.20 Strange attractor of the memristor-based Chua's system (5.30) in $w-x-y$ state space, for parameters $\alpha = 10$, $\beta = 13$, $\gamma = 0.1$, $\zeta = 1.5$, $a = 0.3$, $b = 0.8$, orders $q_1 = q_2 = 0.98$, $q_3 = 0.99$, $q_4 = 0.97$, initial conditions: $x(0) = 0.8$, $y(0) = 0.05$, $z(0) = 0.007$, $w(0) = 0.6$ and simulation time $T_{sim} = 100\,s$.

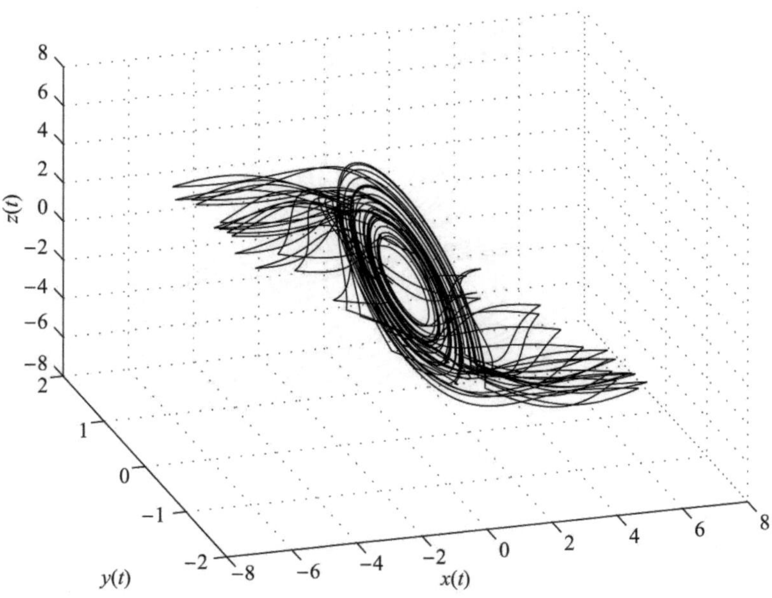

Fig. 5.21 Strange attractor of the memristor-based Chua's system (5.30) in $x-y-z$ state space, for parameters $\alpha = 10$, $\beta = 13$, $\gamma = 0.1$, $\zeta = 1.5$, $a = 0.3$, $b = 0.8$, orders $q_1 = q_2 = 0.98$, $q_3 = 0.99$, $q_4 = 0.97$, initial conditions: $x(0) = 0.8$, $y(0) = 0.05$, $z(0) = 0.007$, $w(0) = 0.6$ and simulation time $T_{sim} = 100\,s$.

$$\lambda^{392} + 4\lambda^{294} + \frac{\lambda^{293}}{10} - 7\lambda^{196} + \frac{67\lambda^{195}}{5} + \frac{383\lambda^{97}}{10} = 0.$$

Both above characteristic equations are polynomials of very high order and it is difficult to find the roots of such polynomials. For the system to remain chaotic, there should be at least one root λ in the unstable region, which means that $|\arg(\lambda)| < \pi/200$.

Because of roots calculation problem, we can predict one unstable eigevalue and assume that the stability condition for chaos is satisfied. It can be indirectly proved via the double-scroll attractor, which can be observed in Fig. 5.20.

In Fig. 5.22 and Fig. 5.23 are depicted the attractors of the memristor-based Chua's system (5.30) for parameters (5.35), $a = 0.3$, $b = 0.8$, orders $q_1 = q_2 = 0.98$, $q_3 = 0.99$, $q_4 = 0.97$, initial conditions: $x(0) = 0.8, y(0) = 0.05, z(0) = 0.007, w(0) = 0.6$ and simulation time $T_{sim} = 100\,s$, projected onto $y - w$, and $z - w$ planes, respectively. These strange attractors also indirectly confirm that the system is chaotic.

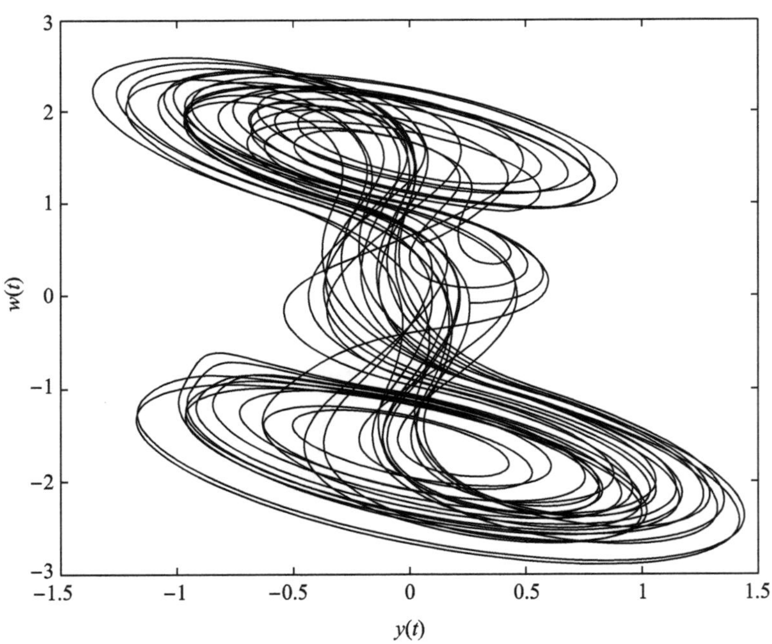

Fig. 5.22 Strange attractor of the memristor-based Chua's system projected onto $y - w$ plane.

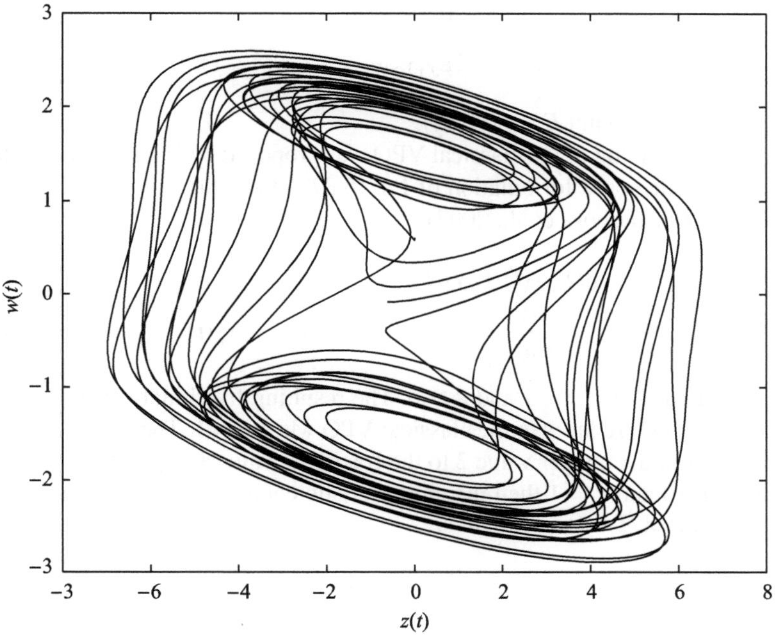

Fig. 5.23 Strange attractor of the memristor-based Chua's system projected onto $z - w$ plane.

5.3 Fractional-Order Van der Pol Oscillator

The Van der Pol oscillator (VPO) represents a nonlinear system with an interesting behavior that exhibits naturally in several applications. It has been used for study and design of many models including biological phenomena, such as the heartbeat, neurons, acoustic models, radiation of mobile phones, and as a model of electrical oscillators (implemented with a tunnel diode, memristor or operating amplifier).

The VPO model was used by Van der Pol in 1920 to study oscillations in vacuum tube circuits. In the standard form, it is given by a nonlinear differential equation of type:

$$y''(t) + \varepsilon(y(t)^2 - 1)y'(t) + y(t) = 0, \tag{5.36}$$

where ε is the control parameter. Equation (5.36) can be rewritten into its state-space representation as follows:

$$\begin{aligned} \frac{dy_1}{dt} &= y_2(t), \\ \frac{dy_2}{dt} &= -y_1(t) - \varepsilon(y_1^2(t) - 1)y_2(t), \end{aligned} \tag{5.37}$$

with an equilibrium point in origin. The Jacobian matrix of the system (5.37) is

$$\mathbf{J} = \begin{bmatrix} 0 & 1 \\ -1-2\varepsilon y_1^* y_2^* & -\varepsilon(y_1^{*2}-1) \end{bmatrix} \tag{5.38}$$

for the equilibrium point $E^* = (y_1^*, y_2^*)$.

A modified version of the classical VPO was proposed by fractional derivative of order q in a state space formulation of Eq. (5.37). It has the following form (Chen and Chen, 2008; Barbosa et al., 2007):

$$\begin{aligned} {}_0D_t^q y_1(t) &= y_2(t), \\ \frac{dy_2}{dt} &= -y_1(t) - \varepsilon(y_1^2(t)-1)y_2(t), \end{aligned} \tag{5.39}$$

where the order is $0 < q < 1$ and $\varepsilon > 0$. The resulting fractional-order Van der Pol oscillator (FrVPO) reduces to the classical VPO when $q = 1$. The total system order is changed from the integer value 2 to the fractional value $1+q < 2$. If we consider $q = 0.9 = 9/10$ and $\varepsilon = 1$ then the characteristic equation of the system (5.39) for $\gamma = 1/10$ is $\det(\lambda^{\gamma}\mathbf{I} - \mathbf{J}) = 0$, that is,

$$\lambda^{19} + \lambda^{9} + 1 = 0.$$

All equation roots λ_i satisfy the condition $|\arg(\lambda_i)| > \pi/20$ for $(i = 1, 2, \ldots, 19)$ and therefore the system is stable. Detailed analysis of the fractional-order Van der Pol system for various system orders has been made in (Barbosa et al., 2007; Ge and Hsu, 2007). This analysis may be useful for a better understanding and control of such system.

In Fig. 5.24 is depicted the limit cycle in the phase plane of the fractional-order Van der Pol oscillator (5.39) for simulation time $T_{sim} = 30\,s$ and time step $h = 0.005$.

Let us consider the modified version of the FrVPO in the following form:

$$\begin{aligned} {}_0D_t^{q_1} y_1(t) &= y_2(t), \\ {}_0D_t^{q_2} y_2(t) &= -y_1(t) - \varepsilon(y_1^2(t)-1)y_2(t), \end{aligned} \tag{5.40}$$

where q_1 and q_2 are orders $(0 < q_{1,2} < 2)$ and $\varepsilon > 0$. If we consider $q_1 = q_2 \equiv q$ in (5.40), we obtain a commensurate-order system. The characteristic equation of the commensurate-order system (5.40) is $\det(\lambda^{q}\mathbf{I} - \mathbf{J}) = 0$ and the stability condition is $|\arg(\mathrm{eig}(\mathbf{J}))| > q\pi/2$.

For simulation purpose, we derived a numerical solution of the FrVPO, obtained by using the relations (2.53) and (2.54), which has the following form:

$$\begin{aligned} y_1(t_k) &= y_2(t_{k-1})h^{q_1} - \sum_{j=v}^{k} c_j^{(q_1)} y_1(t_{k-j}), \\ y_2(t_k) &= \left(-y_1(t_k) - \varepsilon(y_1^2(t_k)-1)y_2(t_{k-1})\right)h^{q_2} - \sum_{j=v}^{k} c_j^{(q_2)} y_2(t_{k-j}), \end{aligned} \tag{5.41}$$

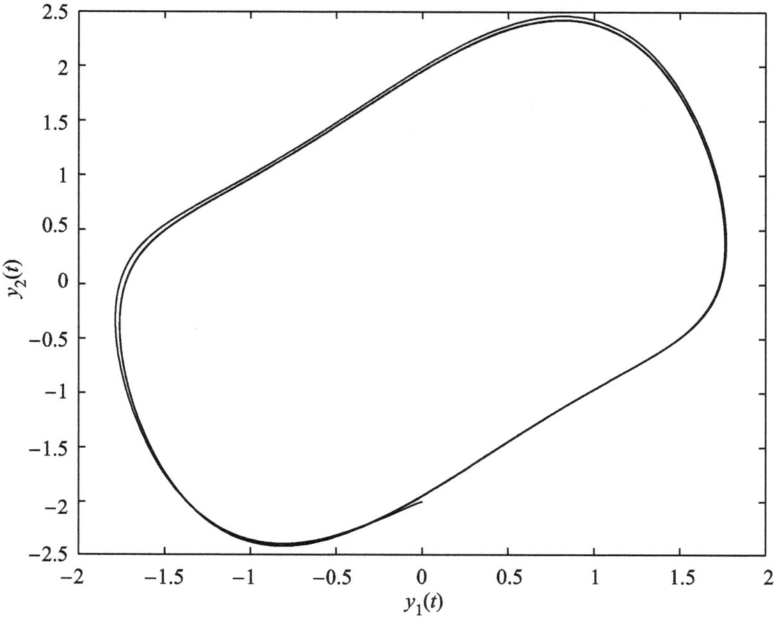

Fig. 5.24 Limit cycle in phase plane $y_1 - y_2$ for FrVPO with fractional order $q = 0.9$, parameter $\varepsilon = 1$, and initial conditions $(y_1(0), y_2(0)) = (0, -2)$.

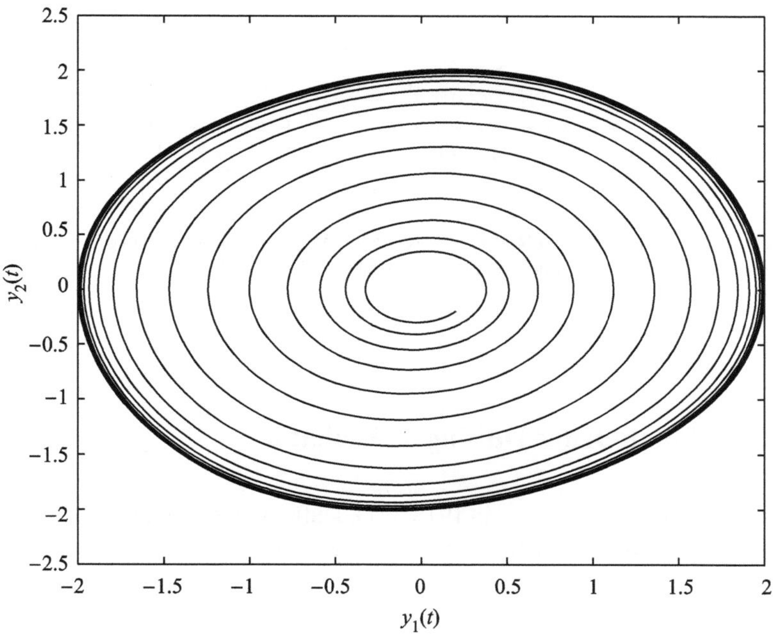

Fig. 5.25 Oscillation in phase plane $y_1 - y_2$ for FrVPO with integer-orders $q_1 = q_2 = 1.0$, parameter $\varepsilon = 0.1$, and initial conditions $(y_1(0), y_2(0)) = (0.2, -0.2)$.

where T_{sim} is the simulation time, $k = 1,2,3,\ldots,N$, for $N = [T_{sim}/h]$, and $(y_1(0), y_2(0))$ is the start point (initial conditions). The binomial coefficients $c_j^{(q_i)}$, $\forall i$ are calculated according to relation (2.54).

In Fig. 5.25 and Fig. 5.26 are depicted the oscillations in the phase plane of the fractional-order Van der Pol oscillator (5.39) for various orders q_1 and q_2, parameter ε, simulation time $T_{sim} = 60\,s$ and time step $h = 0.005$.

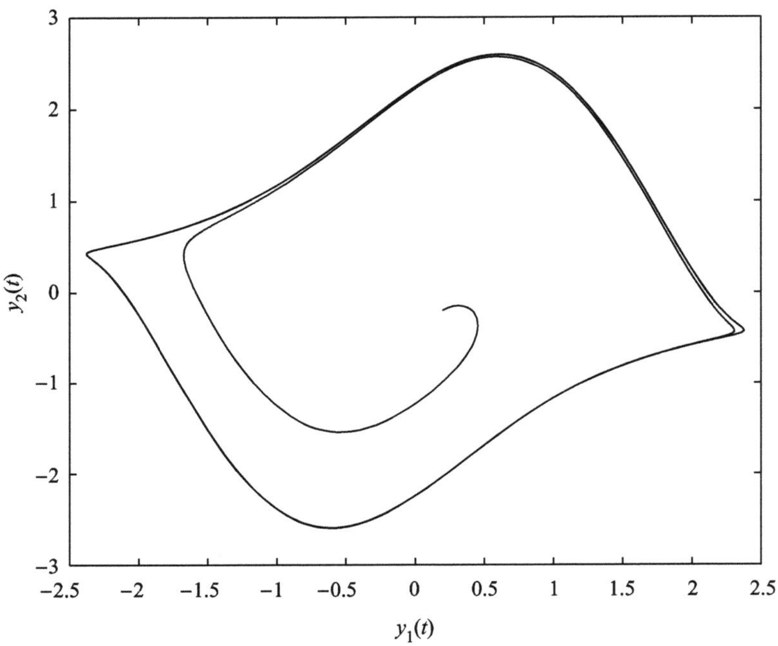

Fig. 5.26 Limit cycle in phase plane $y_1 - y_2$ for FrVPO with fractional-orders $q_1 = 1.2,\ q_2 = 0.8$, parameter $\varepsilon = 1$, and initial conditions $(y_1(0), y_2(0)) = (0.2, -0.2)$.

5.4 Fractional-Order Duffing's Oscillator

Duffing's oscillator, introduced in 1918 by G. Duffing, with negative linear stiffness, damping and periodic excitation is often written in the form

$$x''(t) - x(t) + \alpha x'(t) + x^3(t) = \delta \cos(\omega t). \tag{5.42}$$

Equation (5.42) can be extended to the complex domain in order to study strange attractors and chaotic bahavior of forced vibrations of industrial machinery. The periodically forced complex Duffing's oscillators have the form

$$z''(t) - z(t) + \alpha z'(t) + \varepsilon z|z^2(t)| = \gamma' \cos(\omega t), \tag{5.43}$$

where $\gamma' = \sqrt{2}\gamma e^{j\pi/4}$, γ, α, ω are positive parameters, $z = x + jy$ is a complex function. Equation (5.43) can be reduced to the famous Duffing's oscillator (5.42) when $z = x, (y = 0)$ and $\varepsilon = 1$. When we substitute $z = x + jy$ into Eq. (5.43), we get a system of two coupled nonlinear second-order differential equations (Gao and Yu, 2005):

$$\begin{aligned} x''(t) - x(t) + \alpha x'(t) + \varepsilon x(t)(x^2(t) + y^2(t)) &= \gamma \cos(\omega t), \\ y''(t) - y(t) + \alpha y'(t) + \varepsilon y(t)(x^2(t) + y^2(t)) &= \gamma \cos(\omega t). \end{aligned} \tag{5.44}$$

To get the fractional-order Duffing's system, Equation (5.42) can be rewritten as a system of the first-order autonomous differential equations in the form:

$$\begin{aligned} \frac{x(t)}{dt} &= y(t), \\ \frac{y(t)}{dt} &= x(t) - x^3(t) - \alpha y(t) + \delta \cos(\omega t). \end{aligned} \tag{5.45}$$

Here, the conventional derivatives in Eqs. (5.45) are replaced by the fractional derivatives as follows:

$$\begin{aligned} {}_0D_t^{q_1} x(t) &= y(t), \\ {}_0D_t^{q_2} y(t) &= x(t) - x^3(t) - \alpha y(t) + \delta \cos(\omega t), \end{aligned} \tag{5.46}$$

where q_1, q_2 are two fractional orders and α, δ, ω are the system parameters.

The Jacobian matrix of the Duffing's system is

$$\mathbf{J} = \begin{bmatrix} 0 & 1 \\ 1 - 3x^* & -\alpha \end{bmatrix} \tag{5.47}$$

for the equilibrium point $E^* = (x^*, y^*)$. The characteristic equation of the linearized incommensurate-order system (5.46) for $\gamma = 1/m$ is $\det(\lambda^\gamma \mathbf{I} - \mathbf{J}) = 0$, where m is the LCM of the denominators u_i, if we set $q_i = v_i/u_i$, $v_i, u_i \in Z^+$ for $i = 1,2$. The stability condition is $|\arg(\lambda_i)| > \gamma\pi/2$ for all roots λ_i of the characteristic equation.

A numerical solution of the fractional-order Duffing's system (5.46), obtained by using the relations (2.53) and (2.54), has the following form:

$$\begin{aligned} x(t_k) &= y(t_{k-1}) h^{q_1} - \sum_{j=v}^{k} c_j^{(q_1)} x(t_{k-j}), \\ y(t_k) &= \left(x(t_k) - x^3(t_k) - \alpha y(t_{k-1}) + \delta \cos(\omega t_k)\right) h^{q_2} - \sum_{j=v}^{k} c_j^{(q_2)} y(t_{k-j}), \end{aligned} \tag{5.48}$$

where T_{sim} is the simulation time, $k = 1, 2, 3, \ldots, N$, for $N = [T_{sim}/h]$, and ($x(0)$, $y(0)$) is the start point (initial conditions).

Let us investigate the integer-order Duffing's system (5.45) with parameters $\alpha = 0.15$, $\delta = 0.3$, $\omega = 1$. This system has three fixed points (equilibria): $E_1 = (1.07288371;0)$, $E_2 = (-0.90615851;0)$, and $E_3 = (-0.16672520;0)$ and their stability can be studied by computing the corresponding eigenvalues. For equilibrium E_1 we obtain the eigenvalues $\lambda_{1,2} = -0.0750 \pm 1.487624j$, for E_2 we get $\lambda_1 \approx 1.8547928$, $\lambda_2 \approx -2.004792873$ and for E_3 we have $\lambda_1 \approx 1.1521106$, $\lambda_2 \approx -1.3021106$. The eigenvalues λ_1 and λ_2 of the equilibrium points E_2 and E_3 are saddle points which satisfy the stability condition for chaotic behavior.

In Fig. 5.27 is depicted chaotic attractor of the integer-order Duffing's system (5.45) for the following parameters $\alpha = 0.15$, $\delta = 0.3$, $\omega = 1$ with initial conditions $(x(0), y(0)) = (0.21, 0.13)$ for simulation time $T_{sim} = 200\,s$ and time step $h = 0.005$.

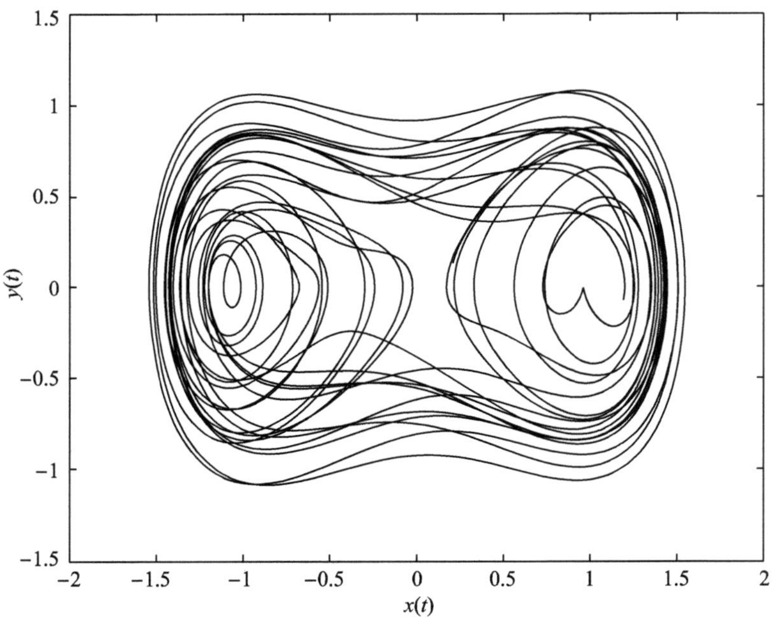

Fig. 5.27 Phase trajectory (attractor) in plane $x - y$ for the integer-order Duffing's system (5.45) with parameters $\alpha = 0.15$, $\delta = 0.3$, $\omega = 1$, and initial conditions $(x(0), y(0)) = (0.21, 0.13)$.

When we assume commensurate orders $q_1 = q_2 = 0.95$ and parameters $\alpha = 0.5$, $\delta = 1.3$, $\omega = 1$ in system (5.46), we obtain a stable limit cycle. All roots of the characteristic equation satisfy the stability condition.

In Fig. 5.28 is depicted the limit cycle of the fractional-order Duffing's system (5.46) for the following parameters $\alpha = 0.5$, $\delta = 1.3$, $\omega = 1$, derivative orders $q_1 = q_2 = 0.95$ with initial conditions $(x(0), y(0)) = (1.0, 1.0)$ for simulation time $T_{sim} = 100\,s$ and time step $h = 0.005$.

In addition, we consider an incommensurate-order system (5.46) with parameters $\alpha = 0.15$, $\delta = 0.3$, $\omega = 1$ and orders $q_1 = 0.9 = 9/10$, and $q_2 = 1.0 = 10/10$. The system has three equilibria and we should investigate the stability of all equilibrium

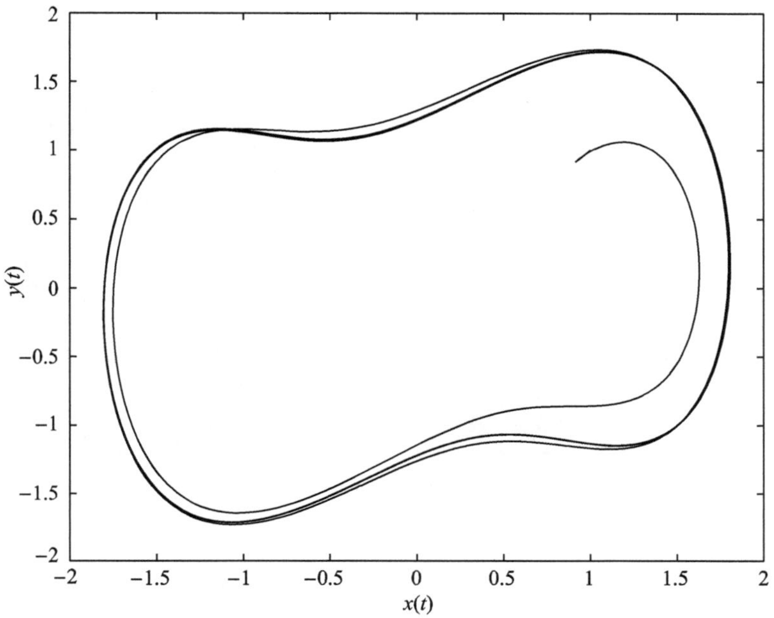

Fig. 5.28 Phase trajectory (limit cycle) in plane $x-y$ for the fractional-order Duffing's system (5.46) with parameters $\alpha = 0.5$, $\delta = 1.3$, $\omega = 1$, derivative orders $q_1 = q_2 = 0.95$, and initial conditions $(x(0), y(0)) = (1.0, 1.0)$.

points. Because of the system parameters, the equilibrium points are the same as in the case of integer-order system. For the equilibrium E_1 the characteristic equation of linearized system is

$$\lambda^{19} + 3/20\lambda^{9} - 3.7184755 = 0,$$

and it has one unstable root $\lambda \approx 1.0673$ because $|\arg(\lambda)| < \pi/20$. For the equilibrium E_2 the characteristic equation of the linearized system is

$$\lambda^{19} + 3/20\lambda^{9} - 1.5001756 = 0,$$

and it has one unstable root $\lambda \approx 1.0151$ because $|\arg(\lambda)| < \pi/20$. Both equilibrium points are unstable nodes. The equilibrium E_3 is a stable focus. The condition to have at least one root in the unstable region in order for the system to be chaotic is satisfied.

In Fig. 5.29 is depicted double scroll attractor of the fractional-order Duffing's system (5.46) for the following parameters $\alpha = 0.15$, $\delta = 0.3$, $\omega = 1$, derivative orders $q_1 = 0.9, q_2 = 1.0$ with initial conditions $(x(0), y(0)) = (0.21, 0.13)$ for simulation time $T_{sim} = 200s$ and time step $h = 0.005$.

An alternative and a bit modified version of the fractional-order Duffing's system and its phase portraits, Poincaré maps, bifurcation diagram and chaotic behavior was

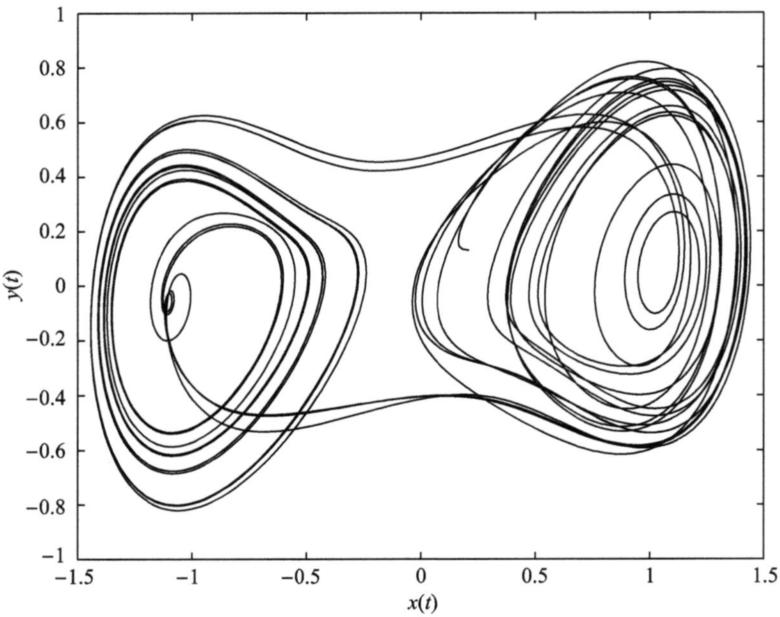

Fig. 5.29 Phase trajectory (attractor) in plane $x-y$ for the fractional-order Duffing's system (5.46) with parameters $\alpha = 0.15$, $\delta = 0.3$, $\omega = 1$, derivative orders $q_1 = 0.9, q_2 = 1.0$, and initial conditions $(x(0), y(0)) = (0.21, 0.13)$.

studied in (Ge and Ou, 2007). The chaotic system reported in the above-mentioned paper considered the Duffing's chaotic system to be an autonomous system with four state variables $x(t)$, $y(t)$, $z(t)$, and $w(t)$ and has the following form:

$$\begin{aligned}
{}_0D_t^{q_1}x(t) &= y(t),\\
{}_0D_t^{q_2}y(t) &= -x(t) - x^3(t) - ay(t) + bz(t),\\
{}_0D_t^{q_3}z(t) &= w(t),\\
{}_0D_t^{q_4}w(t) &= -cz(t) - dz^3(t),
\end{aligned} \tag{5.49}$$

where a, b, c are constant parameters of the system and q_1, q_2, q_3 and q_4 are fractional-order numbers. Usually, the system parameter b is allowed to be variable.

Chaos was found in system (5.49) for the lowest total order of the system 3.8 (Ge and Ou, 2007).

5.5 Fractional-Order Lorenz's System

The Lorenz oscillator is a three-dimensional dynamical system that exhibits chaotic flow. The Lorenz attractor was named after Edward N. Lorenz, who derived it from the simplified equations of convection rolls arising in the equations of the atmosphere in 1963. He for the first time used the term "butterfly effect", which in chaos theory means sensitive dependence on initial conditions. Lorenz wrote a paper in 1979 entitled "Predictability: Does the Flap of a Butterfly's Wings in Brazil Set Off a Tornado in Texas?" Small variations of the initial condition of a dynamical system may produce large variations in the long-term behavior of the system. The phrase refers to the idea that a butterfly's wings might create tiny changes in the atmosphere that may ultimately alter the path of a tornado or delay, accelerate or even prevent the occurrence of a tornado in a certain location. The flapping wing represents a small change in the initial condition of the system, which causes a chain of events leading to large-scale alterations of events.

Lorenz's chaotic system is desribed by

$$\begin{aligned}
\frac{dx(t)}{dt} &= \sigma(y(t)-x(t)),\\
\frac{dy(t)}{dt} &= x(t)(\rho-z(t))-y(t),\\
\frac{dz(t)}{dt} &= x(t)y(t)-\beta z(t),
\end{aligned} \tag{5.50}$$

where σ is called the Prandtl number and ρ is called the Rayleigh number. All $\sigma,\rho,\beta>0$, but usually $\sigma=10$, $\beta=8/3$ and ρ is varied. The system exhibits chaotic behavior for $\rho=28$ and displays orbits for other values.

Lorenz's system has three equilibria, where one is obviously in origin $E_1=(0;0;0)$ and the other two are: $E_2=(\sqrt{(\beta\rho-\beta)};\ \sqrt{(\beta\rho-\beta)};\ \rho-1)$, $E_3=(-\sqrt{(\beta\rho-\beta)};\ -\sqrt{(\beta\rho-\beta)};\ \rho-1)$. The Jacobian matrix of Lorenz's system (5.50) at the equilibrium point $E^*=(x^*,y^*,z^*)$ is given by

$$\mathbf{J}=\begin{bmatrix} -\sigma & \sigma & 0\\ \rho-z^* & -1 & -x^*\\ y^* & x^* & -\beta \end{bmatrix}. \tag{5.51}$$

The equilibrium points of the system with the above parameters are: $E_1=(0;0;0)$, $E_2=(8.4853;8.4853;27)$, and $E_3=(-8.4853;-8.4853;27)$.

The fractional-order Lorenz's system is described as (e.g. (Li and Yan, 2007)):

$$\begin{aligned}
{}_0D_t^{q_1}x(t) &= \sigma(y(t)-x(t)),\\
{}_0D_t^{q_2}y(t) &= x(t)(\rho-z(t))-y(t),\\
{}_0D_t^{q_3}z(t) &= x(t)y(t)-\beta z(t),
\end{aligned} \tag{5.52}$$

where q_1, q_2, and q_3 are derivative orders.

The numerical solution of the fractional-order Lorenz's system has the following form:

$$x(t_k)=\left(\sigma(y(t_{k-1})-x(t_{k-1}))\right)h^{q_1}-\sum_{j=v}^{k}c_j^{(q_1)}x(t_{k-j}),$$

$$y(t_k) = (x(t_k)(\rho - z(t_{k-1})) - y(t_{k-1}))\,h^{q_2} - \sum_{j=v}^{k} c_j^{(q_2)} y(t_{k-j}), \tag{5.53}$$

$$z(t_k) = (x(t_k)y(t_k) - \beta z(t_{k-1}))\,h^{q_3} - \sum_{j=v}^{k} c_j^{(q_3)} z(t_{k-j}),$$

where T_{sim} is the simulation time, $k = 1,2,3,\dots,N$, for $N = [T_{sim}/h]$, and ($x(0)$, $y(0)$, $z(0)$) is the start point (initial conditions). The binomial coefficients $c_j^{(q_i)}$, $\forall i$ are calculated according to the relation (2.54).

To determine a minimal order for which the Lorenz system is chaotic with the parameters $(\sigma,\rho,\beta) = (10,28,8/3)$, we can use the relation (4.42). In this case the minimal commensurate order is $q > 0.9941$, if we consider $q_1 = q_2 = q_3 \equiv q$.

Let us set $q_1 = q_2 = q_3 = 0.995$, the fractional-order Lorenz's system (5.52) has a chaotic attractor as depicted in Fig. 5.30 – Fig. 5.32.

In Fig. 5.30 – Fig. 5.32 are depicted the simulation results of the Lorenz system (5.52) for the following parameters: $\sigma = 10, \rho = 28, \beta = 8/3$, orders $q_1 = q_2 = q_3 = 0.995$ and computational time $100\,s$ for time step $h = 0.005$.

In case of the incommensurate orders (q_1, q_2, q_3) of the system (5.52), the stability at the equilibrium can be investigated via characteristic equation $\det(\lambda^{\gamma}\mathbf{I} - \mathbf{J}) = 0$, for $\gamma = 1/m$, where m is the LCM of the denominators u_i, if we set $q_i = v_i/u_i$, $v_i, u_i \in Z^+$ for $i = 1,2,3$, and the stability condition $|\arg(\lambda)| > \gamma\pi/2$.

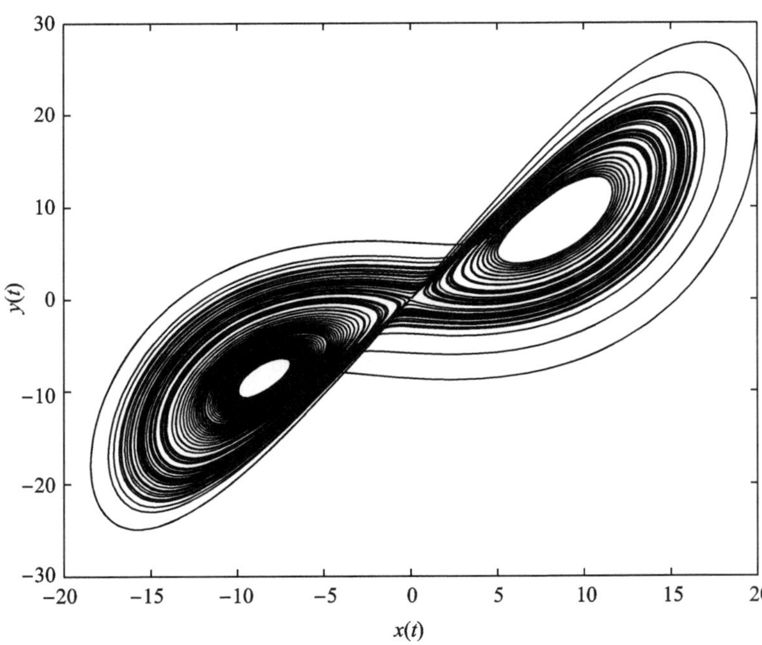

Fig. 5.30 Simulation result of the Lorenz system (5.52) in $x-y$ plane for parameters: $\sigma = 10, \rho = 28, \beta = 8/3$, orders $q_1 = q_2 = q_3 = 0.995$, and initial conditions $(x(0), y(0), z(0)) = (0.1, 0.1, 0.1)$.

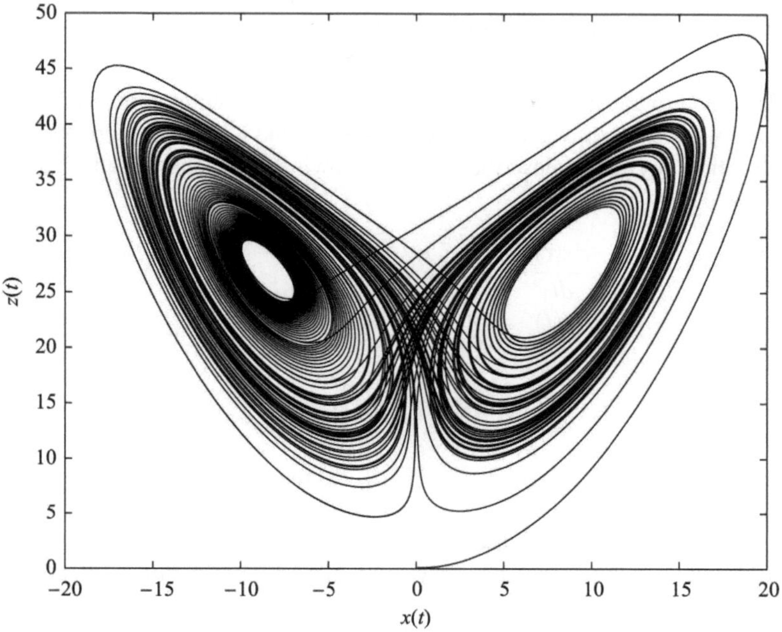

Fig. 5.31 Simulation result of the Lorenz system (5.52) in $x-z$ plane for parameters: $\sigma = 10, \rho = 28, \beta = 8/3$, orders $q_1 = q_2 = q_3 = 0.995$, and initial conditions $(x(0), y(0), z(0)) = (0.1, 0.1, 0.1)$.

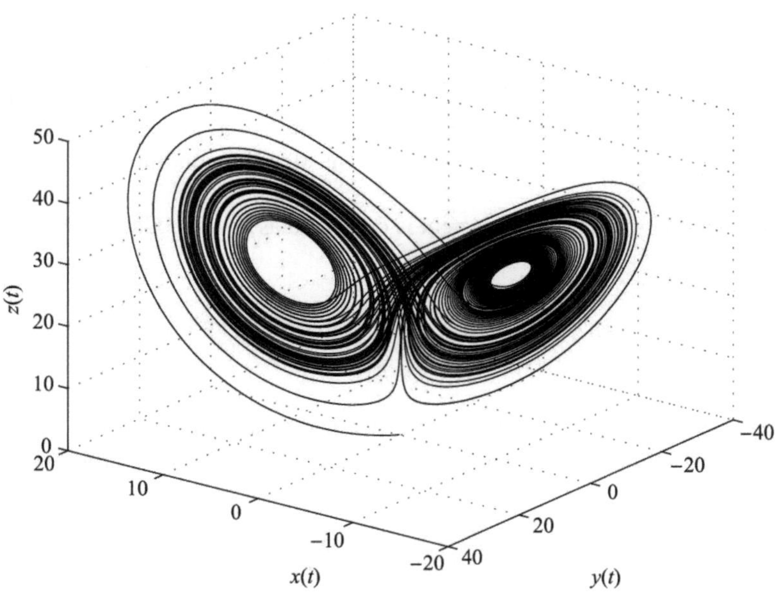

Fig. 5.32 Simulation result of the Lorenz system (5.52) in state space for parameters: $\sigma = 10, \rho = 28, \beta = 8/3$, orders $q_1 = q_2 = q_3 = 0.995$, and initial conditions $(x(0), y(0), z(0)) = (0.1, 0.1, 0.1)$.

5.6 Fractional-Order Chen's System

In 1999, Chen found another a simple three-dimensional autonomous system, which is not topologically equivalent to Lorenz's system and which has a chaotic attractor as well. Chen's system is described by the following equations (Lu and Chen, 2002; Zhou et al., 2004):

$$\begin{aligned}\frac{dx(t)}{dt} &= a(y(t)-x(t)),\\ \frac{dy(t)}{dt} &= (c-a)x(t)-x(t)z(t)+cy(t),\\ \frac{dz(t)}{dt} &= x(t)y(t)-bz(t),\end{aligned} \tag{5.54}$$

where $(a,b,c) \in R^3$. When $(a,b,c) = (35,3,28)$ the chaotic attractor exists.

The equilibrium points of the system with the above parameters are: $E_1 = (0;0;0)$, $E_2 = (7.9373;7.9373;21)$, and $E_3 = (-7.9373;-7.9373;21)$.

The Jacobian matrix of Chen's system (5.54) at the equilibrium point $E^* = (x^*,y^*,z^*)$ is given by

$$\mathbf{J} = \begin{bmatrix} -a & a & 0 \\ c-a-z^* & c & -x^* \\ y^* & x^* & -\beta \end{bmatrix}. \tag{5.55}$$

For the equilibrium E_1 we obtain the eigenvalues $\lambda_1 = -3$, $\lambda_2 \approx 23.8359$, and $\lambda_3 \approx -30.8359$, for E_2 we get $\lambda_1 \approx -18.4280$, and $\lambda_{2,3} \approx 4.2140 \pm 14.8846j$, and for E_3 we have $\lambda_1 \approx -18.4280$, $\lambda_{2,3} \approx 4.2140 \pm 14.8846j$. The eigenvalues λ_1, λ_2 and λ_3 show that the equilibrium E_1 is a saddle point, the equilibria E_2 and E_3 are saddle-focus points. All of them satisfy the stability condition to keep chaotic behavior.

The fractional-order Chen's system is described as (Lu and Chen, 2006):

$$\begin{aligned}{}_0D_t^{q_1}x(t) &= a(y(t)-x(t)),\\ {}_0D_t^{q_2}y(t) &= (c-a)x(t)-x(t)z(t)+cy(t),\\ {}_0D_t^{q_3}z(t) &= x(t)y(t)-bz(t),\end{aligned} \tag{5.56}$$

where $0 < q_1,q_2,q_3 \leq 1$, its total order is denoted by $\bar{q} = (q_1,q_2,q_3)$.

Numerical solution of the fractional-order Chen's system has the following form:

$$\begin{aligned}x(t_k) &= (a(y(t_{k-1})-x(t_{k-1})))\,h^{q_1} - \sum_{j=v}^{k} c_j^{(q_1)} x(t_{k-j}),\\ y(t_k) &= (dx(t_k)-x(t_k)z(t_{k-1})+cy(t_{k-1}))\,h^{q_2} - \sum_{j=v}^{k} c_j^{(q_2)} y(t_{k-j}),\\ z(t_k) &= (x(t_k)y(t_k)-bz(t_{k-1}))\,h^{q_3} - \sum_{j=v}^{k} c_j^{(q_3)} z(t_{k-j}),\end{aligned} \tag{5.57}$$

where $d = (c - a)$, T_{sim} is the simulation time, $k = 1, 2, 3, \ldots, N$, for $N = [T_{sim}/h]$, and $(x(0), y(0), z(0))$ is the start point (initial conditions). The binomial coefficients $c_j^{(q_i)}$, $\forall i$ are calculated according to the relation (2.54).

To determine a minimal order for which the Chen system is chaotic with the parameters $(a, b, c) = (35, 3, 28)$, we can use the relation (4.42). In this case the minimal commensurate order is $q > 0.8244$, if we consider $q_1 = q_2 = q_3 \equiv q$.

Let us consider the parameters $(a, b, c, d) = (35, 3, 28, -7)$ and the commensurate orders $q_1 = q_2 = q_3 = 0.9$ in the numerical solution (5.57).

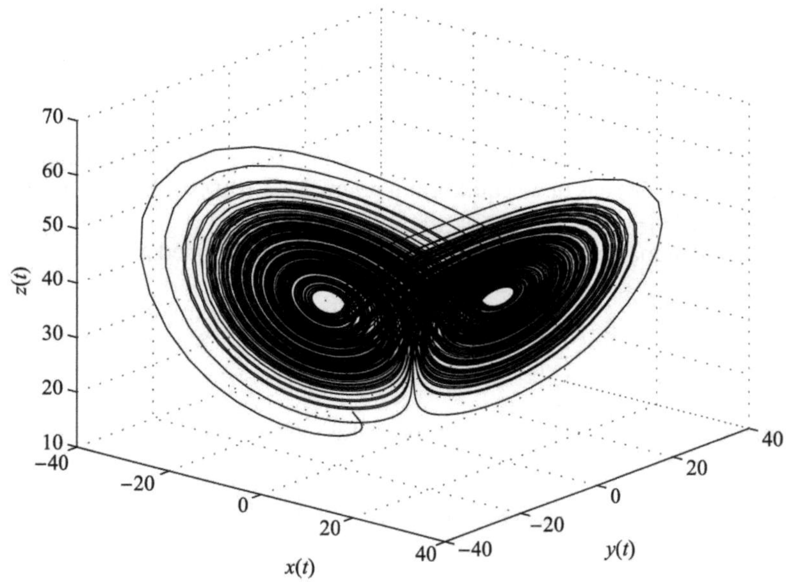

Fig. 5.33 Simulation result of Chen's system (5.56) in state space for parameters: $a = 35, b = 3, c = 28, d = -7$, orders $q_1 = q_2 = q_3 = 0.9$, and initial conditions $(x(0), y(0), z(0)) = (-9, -5, 14)$.

In Fig. 5.33 is depicted the simulation result (double scroll-attractor) of the fractional commensurate-order Chen's system (5.56) computed for simulation time $T_{sim} = 100 s$ and time step $h = 0.005$. For these parameter sets the characteristic equation of the equilibrium points E_2 and E_3 is

$$\lambda^{27} + 10\lambda^{18} + 84\lambda^{9} + 4410 = 0$$

and unstable roots are $\lambda_{1,2} = 1.3417 \pm 0.1944j$, because $|\arg(\lambda_{1,2})| \approx 0.1439 < \pi/2m$, where $m = 10$. These equilibrium points are unstable foci. The equilibrium E_1 is a saddle point connecting two scrolls.

Simulation results of the fractional incommensurate-order Chen's system (5.56) are described in Section 4.3 and depicted in Fig. 4.13, where in Eqs. (5.56) we used the parameters $a = 35, b = 3, c = 28, d = -7$, and orders $q_1 = 0.8,\ q_2 = 1.0$, $q_3 = 0.9$. The stability of such system was investigated as well.

5.7 Fractional-Order Lü's System

The so-called Lü's system is known as a bridge between the Lorenz system and Chen's system. Its fractional version is described as follows (Deng and Li, 2005):

$$\begin{aligned} {}_0D_t^{q_1}x(t) &= a(y(t)-x(t)), \\ {}_0D_t^{q_2}y(t) &= -x(t)z(t)+cy(t), \\ {}_0D_t^{q_3}z(t) &= x(t)y(t)-bz(t), \end{aligned} \tag{5.58}$$

where $0 < q_1, q_2, q_3 \leq 1$, are derivatives orders, and a, b, c are system parameters.

The system (5.58) has three equilibrium points $E_1 = (0;0;0)$, $E_2 = (\sqrt{bc}; \sqrt{bc}; c)$ and $E_3 = (-\sqrt{bc}; -\sqrt{bc}; c)$.

The Jacobian matrix for equilibria $E^* = (x^*, y^*, z^*)$ is defined as

$$\mathbf{J} = \begin{bmatrix} -a & a & 0 \\ -z^* & c & -x^* \\ y^* & x^* & -b \end{bmatrix}. \tag{5.59}$$

Numerical solution of the fractional-order Lü's system (5.58) is given as follows:

$$\begin{aligned} x(t_k) &= (a(y(t_{k-1})-x(t_{k-1})))\,h^{q_1} - \sum_{j=v}^{k} c_j^{(q_1)} x(t_{k-j}), \\ y(t_k) &= (-x(t_k)z(t_{k-1})+cy(t_{k-1}))\,h^{q_2} - \sum_{j=v}^{k} c_j^{(q_2)} y(t_{k-j}), \\ z(t_k) &= (x(t_k)y(t_k)-bz(t_{k-1}))\,h^{q_3} - \sum_{j=v}^{k} c_j^{(q_3)} z(t_{k-j}), \end{aligned} \tag{5.60}$$

where T_{sim} is the simulation time, $k = 1, 2, 3, \ldots, N$, for $N = [T_{sim}/h]$, and ($x(0)$, $y(0)$, $z(0)$) is the start point (initial conditions). The binomial coefficients $c_j^{(q_i)}$, $\forall i$ are calculated according to the relation (2.54).

Let us consider the following parameters $a = 36, b = 3, c = 20$ of the system (5.58). For equilibrium points E_1 we obtain the following eigenvalus of the Jacobian matrix (5.59): $\lambda_1 = -3$, $\lambda_2 = 20$ and $\lambda_3 = -36$. It is a saddle point. For the equilibrium $E_2 = (7.7460; 7.7460; 20)$ we get the eigenvalues $\lambda_1 \approx -22.6516$ and $\lambda_{2,3} \approx 1.8258 \pm 13.6887j$. It is a saddle-focus point. The equilibrium point $E_3 = (-7.7460; -7.7460; 20)$ has the same eigenvalues as the equilibrium E_2. From the above eigenvalues we can determine a minimal commensurate order to keep the system chaotic and it is $q > 0.9156$.

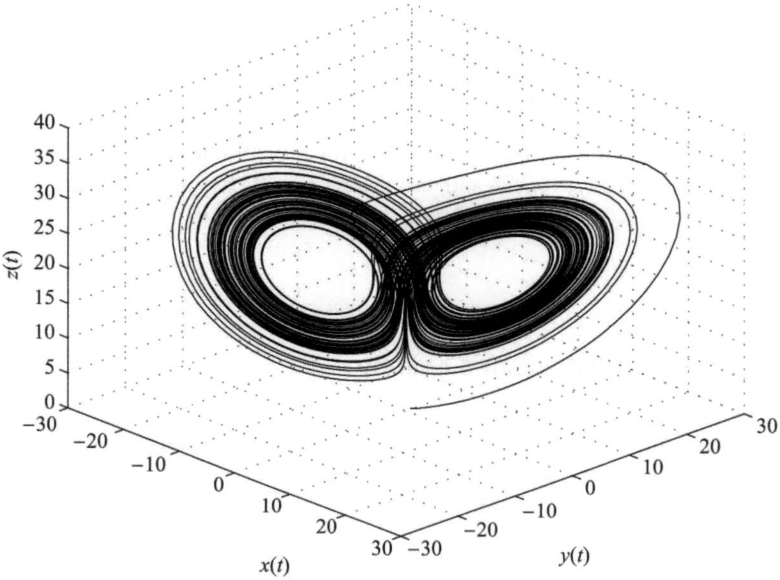

Fig. 5.34 Simulation result of the fractional-order Lü's system (5.58) in state space for parameters $a = 36, b = 3, c = 20$ and orders $q_1 = 0.95, q_2 = 0.95, q_3 = 0.95$.

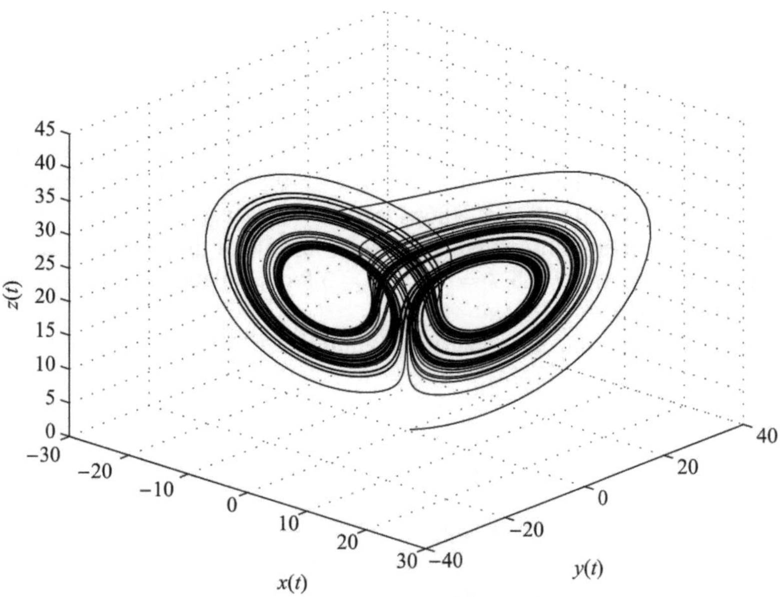

Fig. 5.35 Simulation result of the fractional-order Lü's system (5.58) in state space for parameters $a = 36, b = 3, c = 20$ and orders $q_1 = 0.985, q_2 = 0.99, q_3 = 0.98$.

In Fig. 5.34 is depicted phase trajectory for commensurate derivative orders $q_1 = 0.95, q_2 = 0.95, q_3 = 0.95$ and parameters $a = 36, b = 3, c = 20$ with the initial conditions: $(x(0), y(0), z(0)) = (0.2, 0.5, 0.3)$, for simulation time $90\,s$, and time step $h = 0.005$.

In Fig. 5.35 is depicted phase trajectory for incommensurate derivative orders $q_1 = 0.985, q_2 = 0.99, q_3 = 0.98$ and parameters $a = 36, b = 3, c = 20$ with the initial conditions: $(x(0), y(0), z(0)) = (0.2, 0.5, 0.3)$, for simulation time $90\,s$ and time step $h = 0.005$.

5.8 Fractional-Order Liu's System

A novel three-dimensional autonomous chaotic dynamical system was introduced by C. Liu, L. Liu and T. Liu and reported in literature (Liu et al., 2009). The differential equations that described the system are

$$\begin{aligned}
\frac{dx(t)}{dt} &= -ax(t) - ey^2(t), \\
\frac{dy(t)}{dt} &= by(t) - kx(t)z(t), \\
\frac{dz(t)}{dt} &= -cz(t) + mx(t)y(t),
\end{aligned} \tag{5.61}$$

where $a = e = 1$, $b = 2.5$, $k = m = 4$, $c = 5$ and initial conditions $(0.2, 0, 0.5)$ yield chaotic trajectory.

The system (5.61) has five equilibrium points. Two of them are complex and three are real equilibrium points $E_1 = (0; 0; 0)$, $E_2 = (-0.88388; -0.940150; 0.664786)$, and $E_3 = (-0.88388; 0.940150; -0.664786)$. In the paper (Liu et al., 2009) were calculated different equilibrium points because in spite of the following parameters declaration $a = e = 1$, $b = 2.5$, $k = m = 4$, $c = 5$ for calculations they probably used different parameters.

The corresponding Jacobian matrix for equilibria $E^* = (x^*, y^*, z^*)$ is

$$\mathbf{J} = \begin{bmatrix} -a & -2ey^* & 0 \\ -kz^* & b & -kx^* \\ my^* & mx^* & -c \end{bmatrix}. \tag{5.62}$$

The roots of the characteristic equation evaluated at equilibrium E_1 are $\lambda_1 = -1$, $\lambda_2 = -5$, and $\lambda_3 = 2.5$. It is a saddle point. The eigenvalues of the Jacobian matrix evaluated at equilibrium points E_2 and E_3 are $\lambda_1 \approx -4.387767$, and $\lambda_{2,3} \approx 0.4438837 \pm 3.346383j$. It is a saddle-focus point. Because all eigenvalues are unstable, the condition for chaos is satisfied and chaotic system (5.61) with the above parameters can exhibit chaotic behavior.

Its fractional-order version was described (Gejji and Bhalekar, 2010) and has the form:

$$
\begin{aligned}
{}_0D_t^{q_1}x(t) &= -ax(t) - ey^2(t),\\
{}_0D_t^{q_2}y(t) &= by(t) - kx(t)z(t),\\
{}_0D_t^{q_3}z(t) &= -cz(t) + mx(t)y(t),
\end{aligned} \tag{5.63}
$$

where q_1, q_2, q_3 are derivative orders, the total order is denoted by $\bar{q} = (q_1, q_2, q_3)$.

In case we consider a commensurate order system (5.63) with $q_1 = q_2 = q_3 \equiv q$, a minimal order q for chaotic behavior can be determinned according the condition (4.42) and it is $q > 0.916$. Thus the system deos not show chaotic behavior for $q < 0.916$.

Numerical solution of the fractional-order Liu's system (5.63) is given as follows:

$$
\begin{aligned}
x(t_k) &= \left(-ax(t_{k-1}) - ey^2(t_{k-1})\right) h^{q_1} - \sum_{j=v}^{k} c_j^{(q_1)} x(t_{k-j}),\\
y(t_k) &= \left(by(t_{k-1}) - kx(t_k)z(t_{k-1})\right) h^{q_2} - \sum_{j=v}^{k} c_j^{(q_2)} y(t_{k-j}),\\
z(t_k) &= \left(-cz(t_{k-1}) + mx(t_k)y(t_k)\right) h^{q_3} - \sum_{j=v}^{k} c_j^{(q_3)} z(t_{k-j}),
\end{aligned} \tag{5.64}
$$

where T_{sim} is the simulation time, $k = 1, 2, 3, \ldots, N$, for $N = [T_{sim}/h]$, and $(x(0)$, $y(0)$, $z(0))$ is the start point (initial conditions). The binomial coefficients $c_j^{(q_i)}$, $\forall i$ are calculated according to the relation (2.54).

In Fig. 5.36 are depicted the simulation results of the (integer-order) Liu's system (5.61) for the following parameters: $a = e = 1$, $b = 2.5$, $k = m = 4$, $c = 5$, and computational time $100\,s$, for time step $h = 0.005$.

Consider the commensurate order of the fractional-order Liu's system (5.63) with $q = 0.95$ and parameters $a = e = 1$, $b = 2.5$, $k = m = 4$. The characteristic equation of the linearized system is

$$\lambda^{285} + 3.5\lambda^{190} + 7.5\lambda^{95} + 50 = 0$$

and unstable roots are $\lambda_{1,2} \approx 1.0128 \pm 0.0153j$, because $|\arg(\lambda_{1,2})| = 0.0151 < \pi/2m$, where $m = 100$ (LCM of orders denominator).

In Fig. 5.37 is depicted the simulation result of the (integer-order) Liu's system (5.63) for the following parameters: $a = e = 1$, $b = 2.5$, $k = m = 4$, $c = 5$, orders $q_1 = q_2 = q_3 = 0.95$ and computational time $100s$ for time step $h = 0.005$.

Let us consider the incommensurate order of the fractional-order Liu's system (5.63) with $q_1 = 1.0$, $q_2 = 0.9$, and $q_3 = 0.8$ and parameters $a = e = 1$, $b = 2.5$, $k = m = 4$. The characteristic equation of the linearized system is

$$\lambda^{27} + 5\lambda^{19} - 2.5\lambda^{18} + \lambda^{17} + 5\lambda^{9} + 2.5\lambda^{8} + 50 = 0$$

and unstable roots are $\lambda_{1,2} \approx 1.1224 \pm 0.1770j$, because $|\arg(\lambda_{1,2})| = 0.1565 < \pi/2m$, where $m = 10$.

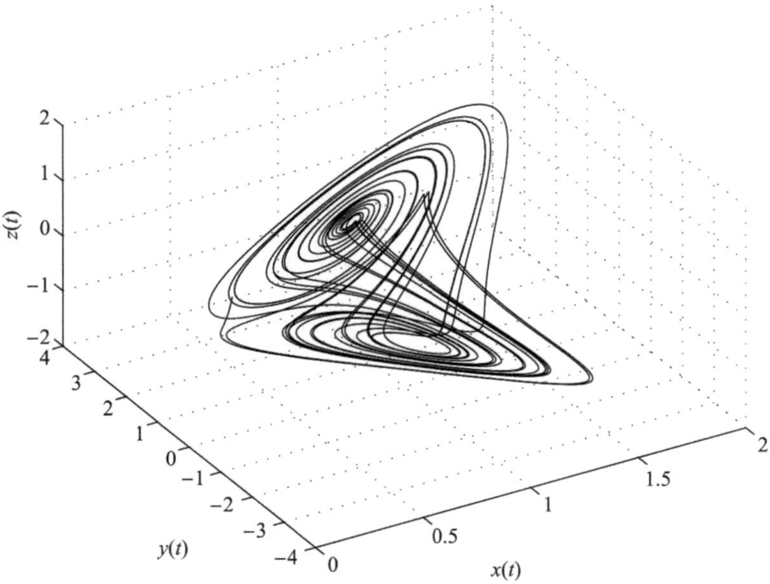

Fig. 5.36 Simulation result of the Liu's system (5.61) in state space for parameters: $a = e = 1$, $b = 2.5$, $k = m = 4$, $c = 5$, and initial conditions $(x(0), y(0), z(0)) = (0.2, 0, 0.5)$.

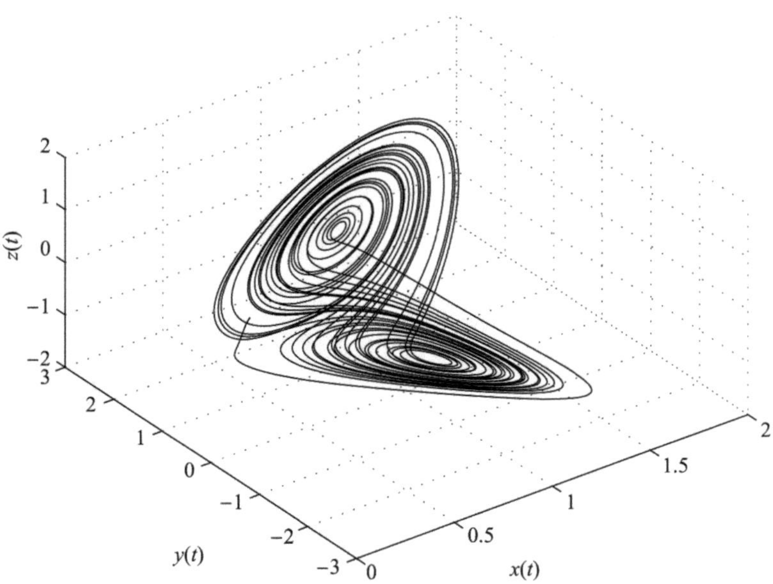

Fig. 5.37 Simulation result of the Liu's system (5.63) in state space for parameters: $a = e = 1$, $b = 2.5$, $k = m = 4$, $c = 5$, orders $q_1 = q_2 = q_3 = 0.95$, and initial conditions $(x(0), y(0), z(0)) = (0.2, 0, 0.5)$.

In Fig. 5.38 is depicted the simulation result of the (fractional-order) Liu's system (5.63) for the following parameters: $a = e = 1$, $b = 2.5$, $k = m = 4$, $c = 5$, orders $q_1 = 1.0$, $q_2 = 0.9$, and $q_3 = 0.8$, and computational time $100\,s$, for time step $h = 0.005$.

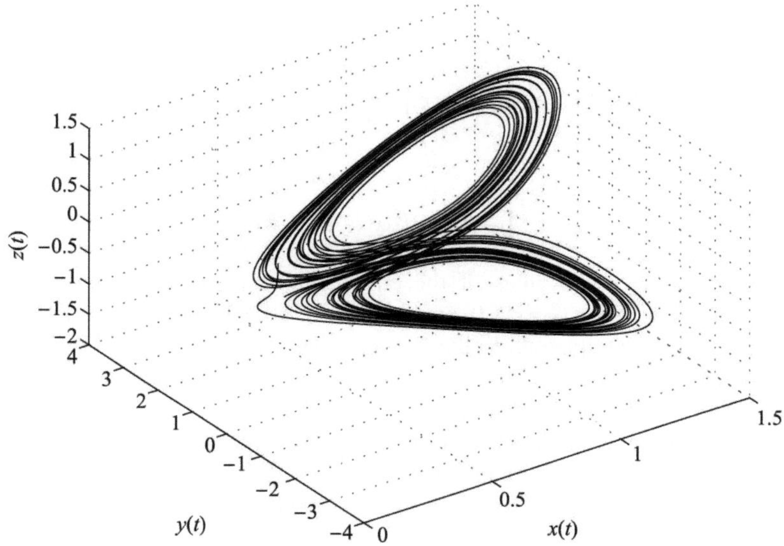

Fig. 5.38 Simulation result of the Liu's system (5.63) in state space for parameters: $a = e = 1$, $b = 2.5$, $k = m = 4$, $c = 5$, orders $q_1 = 1.0$, $q_2 = 0.9$, and $q_3 = 0.8$, and initial conditions $(x(0), y(0), z(0)) = (0.2, 0, 0.5)$.

Note that this fractional-order system has been investigated and described for various orders, where condition for chaotic behavior was cross-validated with the Lyapunov exponent together with the instability measure for each equilibrium point (Gejji and Bhalekar, 2010). Numerical experiments performed in the mentioned paper showed the existence of chaos for a minimum total order of the commensurate-order system 2.76 and in the case of incommensurate-order system it was 2.60. The total order of the system is sometimes also called a minimum effective dimension.

5.9 Fractional-Order Genesio-Tesi's System

The Genesio-Tesi's system is described by the system of equations (Genesio and Tesi, 1992):

$$
\begin{aligned}
\frac{dx(t)}{dt} &= y(t),\\
\frac{dy(t)}{dt} &= z(t),\\
\frac{dz(t)}{dt} &= -\beta_1 x(t) - \beta_2 y(t) - \beta_3 z(t) + \beta_4 x^2(t),
\end{aligned}
\tag{5.65}
$$

where $\beta_1, \beta_2, \beta_3$ and β_4 are system parameters.

Genesio-Tesi's system (5.65) has two equilibrium points $E_1 = (0;0;0)$ and $E_2 = (\beta_1/\beta_4;0;0)$. The Jacobian matrix for equilibria $E^* = (x^*, y^*, z^*)$ is defined as

$$
\mathbf{J} = \begin{bmatrix} 0 & 1 & 0 \\ 0 & 0 & 1 \\ -\beta_1 + 2\beta_4 x^* & -\beta_2 & -\beta_3 \end{bmatrix}. \tag{5.66}
$$

The corresponding eigenvalues of the equilibrium E_1 for the parameters $\beta_1 = 1, \beta_2 = 1.1, \beta_3 = 0.44, \beta_4 = 1.0$ are $\lambda_1 \approx -0.750293$ and $\lambda_{2,3} \approx 0.155146 \pm 1.144002j$. For the equilibrium E_2 they are $\lambda_1 \approx 0.587161$ and $\lambda_{2,3} \approx -0.5135806 \pm 1.199726j$. Both of them are unstable saddle-focus points and therefore the condition for chaotic behavior is satisfied.

In Fig. 5.39 is depicted the simulation result of the (integer-order) Genesio-Tesi's system (5.65) for the following parameters: $\beta_1 = 1$, $\beta_2 = 1.1$, $\beta_3 = 0.44$, $\beta_4 = 1.0$, and computational time $200\,s$, for time step $h = 0.005$.

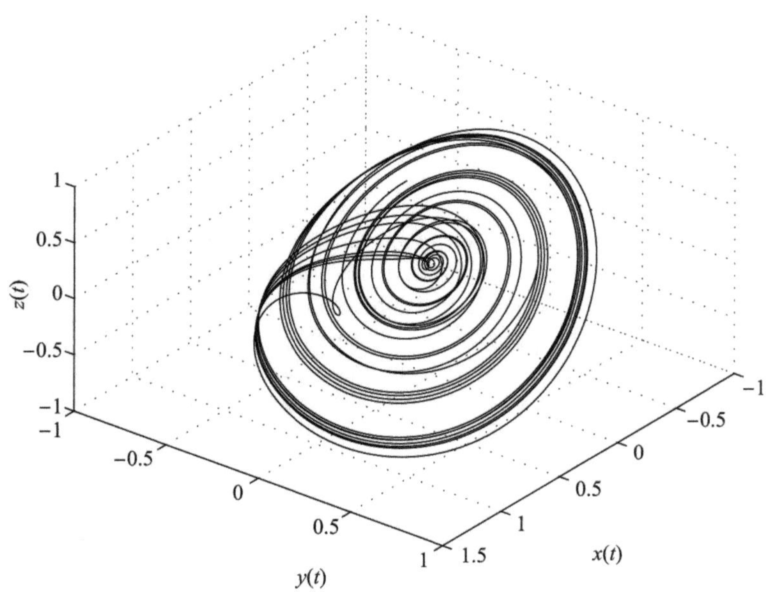

Fig. 5.39 Simulation result of the Genesio-Tesi's system (5.65) in state space for parameters: $\beta_1 = 1$, $\beta_2 = 1.1$, $\beta_3 = 0.44$, $\beta_4 = 1.0$, and initial conditions $(x(0), y(0), z(0)) = (-0.1, 0.5, 0.2)$.

The fractional-order Genesio-Tesi's system is defined as follows (Guo, 2005):

$$\begin{aligned} {}_0D_t^{q_1}x(t) &= y(t), \\ {}_0D_t^{q_2}y(t) &= z(t), \\ {}_0D_t^{q_3}z(t) &= -\beta_1 x(t) - \beta_2 y(t) - \beta_3 z(t) + \beta_4 x^2(t), \end{aligned} \tag{5.67}$$

where $q \in [q_1, q_2, q_3]$ and $0 < q \leq 1$.

Numerical solution of the fractional-order Genesio-Tesi's system (5.67) is given as follows:

$$\begin{aligned} x(t_k) &= (y(t_{k-1}))\, h^{q_1} - \sum_{j=v}^{k} c_j^{(q_1)} x(t_{k-j}), \\ y(t_k) &= (z(t_{k-1}))\, h^{q_2} - \sum_{j=v}^{k} c_j^{(q_2)} y(t_{k-j}), \\ z(t_k) &= \left(-\beta_1 x(t_k) - \beta_2 y(t_k) - \beta_3 z(t_{k-1}) + \beta_4 x(t_k)^2\right) h^{q_3} - \sum_{j=v}^{k} c_j^{(q_3)} z(t_{k-j}), \end{aligned} \tag{5.68}$$

where T_{sim} is the simulation time, $k = 1, 2, 3, \ldots, N$, for $N = [T_{sim}/h]$, and $(x(0), y(0), z(0))$ is the start point (initial conditions). The binomial coefficients $c_j^{(q_i)}$, $\forall i$ are calculated according to the relation (2.54).

Let us consider the commensurate order $q_1 = q_2 = q_3 \equiv q = 0.9$ and system parameters: $\beta_1 = 1$, $\beta_2 = 1.1$, $\beta_3 = 0.15$, $\beta_4 = 1.0$. The characteristic equation of the system (5.67) evaluated at the equilibria E_1 and E_2, respectively, is

$$\lambda^{27} + 3/20\lambda^{18} + 11/10\lambda^{9} \pm 1 = 0$$

and unstable roots for the equilibrium E_1 are $\lambda_{1,2} \approx 1.0100 \pm 0.1525j$, because $|\arg(\lambda_{1,2})| = 0.1499 < \pi/2m$, where $m = 10$ (LCM of orders denominator) and unstable root for the equilibrium E_2 is $\lambda_1 \approx 0.9498$. As shown in (Guo, 2005), the system exhibits chaotic behavior.

Now, consider the incommensurate-order system, where $q_1 = 1.0$, $q_2 = 1.0$, and $q_3 = 0.95$ and system parameters are $\beta_1 = 1.1$, $\beta_2 = 1.1$, $\beta_3 = 0.45$, $\beta_4 = 1.0$. The characteristic equation of the system (5.67) evaluated at the equilibria E_1 and E_2, respectively, is

$$\lambda^{295} + 9/20\lambda^{200} + 11/10\lambda^{100} \pm 1.1 = 0$$

and unstable roots for the equilibrium E_1 are $\lambda_{1,2} \approx 1.0014 \pm 0.0145j$, because $|\arg(\lambda_{1,2})| = 0.0145 < \pi/2m$, where $m = 100$ (LCM of orders denominator) and unstable root for the equilibrium E_2 is $\lambda_1 \approx 0.9952$.

In Fig. 5.40 is depicted the simulation result of the Genesio-Tesi's system (5.67) for the following parameters: $\beta_1 = 1.1$, $\beta_2 = 1.1$, $\beta_3 = 0.45$, $\beta_4 = 1.0$, orders $q_1 = 1.0$, $q_2 = 1.0$, $q_3 = 0.95$ and computational time $200\,s$, for time step $h = 0.005$. As we can see in the figure, the fractional-order Genesio-Tesi's system is chaotic with one scroll attractor.

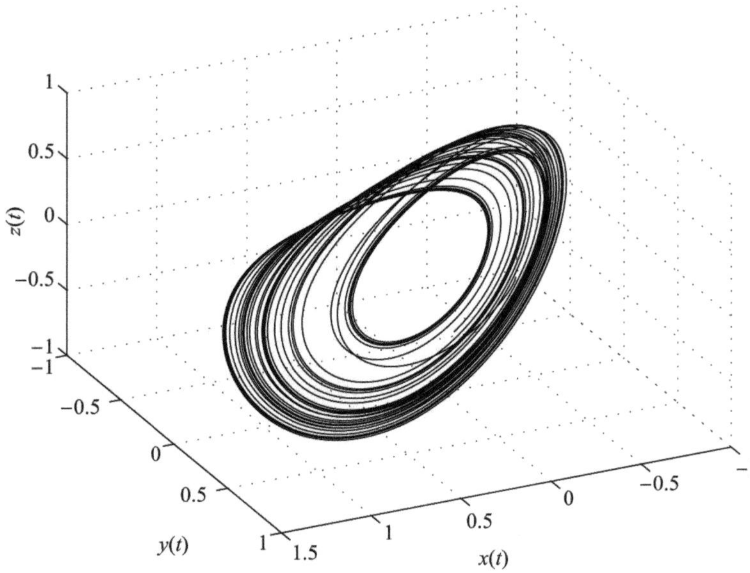

Fig. 5.40 Simulation result of the Genesio-Tesi's system (5.67) in state space for initial conditions $(x(0), y(0), z(0)) = (-0.1, 0.5, 0.2)$.

5.10 Fractional-Order Arneodo's System

Arneodo's system is described by

$$\begin{aligned} \frac{dx(t)}{dt} &= y(t), \\ \frac{dy(t)}{dt} &= z(t), \\ \frac{dz(t)}{dt} &= -\beta_1 x(t) - \beta_2 y(t) - \beta_3 z(t) + \beta_4 x^3(t), \end{aligned} \tag{5.69}$$

where $\beta_1, \beta_2, \beta_3$ and β_4 are constant parameters. This system has three equilibrium points $E_1 = (0;0;0;)$, $E_2 = (\sqrt{(\beta_4\beta_1)}/\beta_4;0;0)$, and $E_3 = (-\sqrt{(\beta_4\beta_1)}/\beta_4;0;0)$.

The Jacobian matrix for equilibria $E^* = (x^*, y^*, z^*)$ is defined as

$$\mathbf{J} = \begin{bmatrix} 0 & 1 & 0 \\ 0 & 0 & 1 \\ -\beta_1 + 3\beta_4 x^{*2} & -\beta_2 & -\beta_3 \end{bmatrix}. \tag{5.70}$$

When $\beta_1 = -5.5$, $\beta_2 = 3.5$, $\beta_3 = 1$ and $\beta_4 = -1$, the system (5.69) equilibrium points are $E_1 = (0;0;0)$, $E_2 = (2.345207;0;0)$, and $E_3 = (-2.345207;0;0)$. Corresponding eigenvalues for equilibrium E_1 are $\lambda_1 = 1$, $\lambda_{2,3} \approx -1 \pm 2.12132j$, and for equilibria E_2 and E_3 the eigenvalues are $\lambda_1 = -2$, $\lambda_{2,3} = 0.5 \pm 2.2912878j$. All equilibria are saddle-focus points. The condition for chaos is satisfied and the system has chaotic attractor shown in Fig. 5.41.

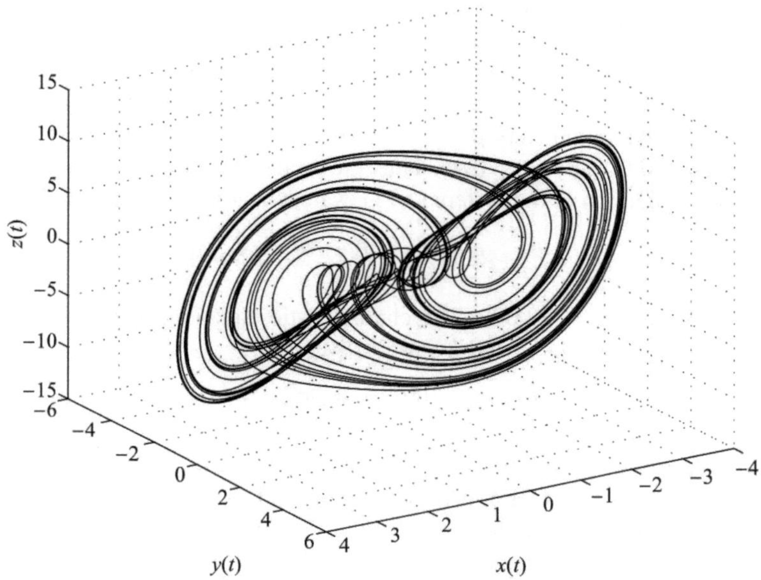

Fig. 5.41 Simulation result of the Arneodo's system (5.69) in state space for initial conditions $(x(0), y(0), z(0)) = (-0.2, 0.5, 0.2)$.

In Fig. 5.41 is depicted the chaotic attractor of Arneodo's system (5.69) for the following parameters: $\beta_1 = -5.5$, $\beta_2 = 3.5$, $\beta_3 = 1$, $\beta_4 = -1.0$, and computational time $200\,s$, for time step $h = 0.005$.

The fractional-order Arneodo's system is defined as follows (Lu, 2005):

$$\begin{aligned} {}_0D_t^{q_1}x(t) &= y(t), \\ {}_0D_t^{q_2}y(t) &= z(t), \\ {}_0D_t^{q_3}z(t) &= -\beta_1 x(t) - \beta_2 y(t) - \beta_3 z(t) + \beta_4 x^3(t), \end{aligned} \tag{5.71}$$

where $q \in [q_1, q_2, q_3]$ and $0 < q \leq 1$. This system is very similar to the Genesio-Tesi's system (5.67) but with the different kind of nonlinearity.

Numerical solution of the fractional-order Arneodo's system (5.71) is given as follows:

$$\begin{aligned} x(t_k) &= (y(t_{k-1}))\,h^{q_1} - \sum_{j=v}^{k} c_j^{(q_1)} x(t_{k-j}), \\ y(t_k) &= (z(t_{k-1}))\,h^{q_2} - \sum_{j=v}^{k} c_j^{(q_2)} y(t_{k-j}), \\ z(t_k) &= \left(-\beta_1 x(t_k) - \beta_2 y(t_k) - \beta_3 z(t_{k-1}) + \beta_4 x(t_k)^3\right) h^{q_3} - \sum_{j=v}^{k} c_j^{(q_3)} z(t_{k-j}), \end{aligned} \tag{5.72}$$

where T_{sim} is the simulation time, $k = 1,2,3,\ldots,N$, for $N = [T_{sim}/h]$, and $(x(0)$, $y(0)$, $z(0))$ is the start point (initial conditions). The binomial coefficients $c_j^{(q_i)}$, $\forall i$ are calculated according to the relation (2.54).

For the above paremeters we are able to determine a minimal commensurate order of the system (5.71), which is $q > 0.86$ in case $q_1 = q_2 = q_3 \equiv q$.

Let us consider the incommensuraty fractional-order Arneodo's system (5.71) with the following paremeters $\beta_1 = -5.5$, $\beta_2 = 3.5$, $\beta_3 = 0.8$, $\beta_4 = -1.0$ and orders $q_1 = q_2 = 0.97$, and $q_3 = 0.96$. Thus the total order of the system is 2.9. The characteristic equation of the system (5.71) evaluated at the equilibrium E_1 is

$$\lambda^{290} + 4/5\lambda^{194} + 7/2\lambda^{97} - 5.5 = 0$$

with unstable root $\lambda_1 \approx 1.0002$. The characteristic equation of the system (5.71) evaluated at the equilibria E_2 and E_3, respectively, is

$$\lambda^{290} + 4/5\lambda^{194} + 7/2\lambda^{97} + 11 = 0$$

and unstable roots are $\lambda_{1,2} \approx 1.0089 \pm 0.0139j$, because $|\arg(\lambda_{1,2})| = 0.0138 < \pi/2m$, where $m = 100$ (LCM of orders denominator).

In Fig. 5.42 and Fig. 5.43 are depicted the simulation results of Arneodo's system (5.71) for the following parameters: $\beta_1 = -5.5$, $\beta_2 = 3.5$, $\beta_3 = 0.8$, $\beta_4 = -1.0$, orders $q_1 = 0.97$, $q_2 = 0.97$, $q_3 = 0.96$, and computational time $200s$, for time step $h = 0.005$.

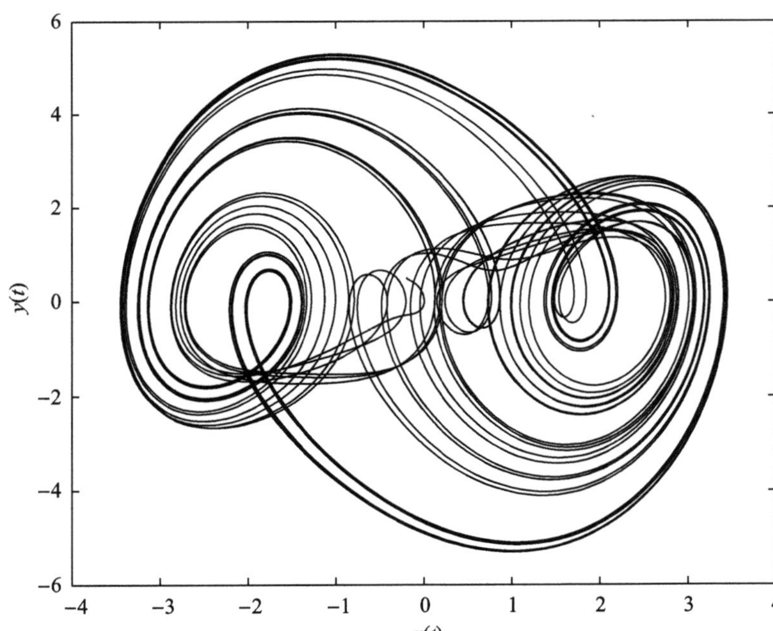

Fig. 5.42 Simulation result of the fractional-order Arneodo's system (5.71) projected onto $x-y$ plane for the initial conditions $(x(0), y(0), z(0)) = (-0.2, 0.5, 0.2)$.

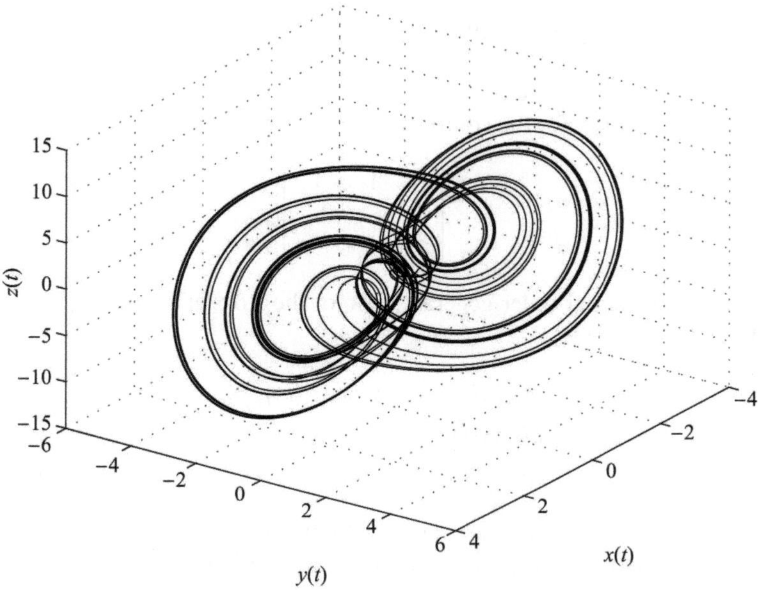

Fig. 5.43 Simulation result of the fractional-order Arneodo's system (5.71) in state space for the initial conditions $(x(0), y(0), z(0)) = (-0.2, 0.5, 0.2)$.

In the paper (Lu, 2005) were performed simulations for various commensurate order q of the fractional-order Arneodo's system (5.71). Those simulations were cross-validated with the Lyapunov exponent. The results showed that the lowest total order of the fractional-order Arneodo's system to yield chaos was 2.1.

5.11 Fractional-Order Rössler's System

Otto Rössler proposed Rössler's system with strange attractor in 1976, but the originally theoretical equations were later found to be useful in modeling equilibrium in chemical reactions. This attractor has only one manifold and can be obtained as a solution of the following equations:

$$\begin{aligned}\frac{dx(t)}{dt} &= -(y(t)+z(t)),\\ \frac{dy(t)}{dt} &= x(t)+ay(t),\\ \frac{dz(t)}{dt} &= b+z(t)\left(x(t)-c\right),\end{aligned} \tag{5.73}$$

where for the parameters $a = 0.2$, $b = 0.2$, $c = 5.7$ this system yields chaotic behavior. This system has two equilibrium points E_1 and E_2 located at

$$E_{1,2} = \left(\frac{c \pm \sqrt{c^2 - 4ab}}{2}; -\frac{c \pm \sqrt{c^2 - 4ab}}{2a}; \frac{c \pm \sqrt{c^2 - 4ab}}{2a} \right).$$

The Jacobian matrix for the equilibria $E^* = (x^*, y^*, z^*)$ is defined as

$$\mathbf{J} = \begin{bmatrix} 0 & -1 & -1 \\ 1 & a & 0 \\ z^* & 0 & x^* - c \end{bmatrix}. \tag{5.74}$$

Consider a fractional-order generalization of the Rössler's system (5.73) as follows (Li and Chen, 2004):

$$\begin{aligned} {}_0D_t^{q1} x(t) &= -(y(t) + z(t)), \\ {}_0D_t^{q2} y(t) &= x(t) + ay(t), \\ {}_0D_t^{q3} z(t) &= b + z(t)\,(x(t) - c)\,, \end{aligned} \tag{5.75}$$

where conventional derivatives are replaced by the fractional ones.

Numerical solution of the fractional-order Rössler's system (5.75) is given as follows:

$$\begin{aligned} x(t_k) &= (-y(t_{k-1}) - z(t_{k-1}))\,h^{q_1} - \sum_{j=v}^{k} c_j^{(q_1)} x(t_{k-j}), \\ y(t_k) &= (x(t_k) + ay(t_{k-1}))\,h^{q_2} - \sum_{j=v}^{k} c_j^{(q_2)} y(t_{k-j}), \\ z(t_k) &= (b + z(t_{k-1})(x(t_k) - c))\,h^{q_3} - \sum_{j=v}^{k} c_j^{(q_3)} z(t_{k-j}), \end{aligned} \tag{5.76}$$

where T_{sim} is the simulation time, $k = 1, 2, 3, \ldots, N$, for $N = [T_{sim}/h]$, and $(x(0), y(0), z(0))$ is the start point (initial conditions). The binomial coefficients $c_j^{(q_i)}$, $\forall i$ are calculated according to the relation (2.54).

Let us consider the system (5.75) for the parameters $a = 0.5$, $b = 0.2$ and $c = 10$. The system has two equilibria: $E_1 = (9.98998; -19.97997; 19.97997)$ and $E_2 = (0.10010; -0.20020; 0.20020)$ and their corresponding eigenvalues are: $\lambda_1 \approx 0.47595$, $\lambda_{2,3} \approx 0.007017 \pm 4.57910j$ for the eqiulibrium E_1 and $\lambda_1 \approx -9.98800$, $\lambda_{2,3} \approx 0.249007 \pm 0.96808j$ for the equilibrium E_2. The equilibrium E_1 is unstable focus-node point and the equibrium E_2 is unstable saddle-focus point. The condition for chaotic behavior is satisfied. With the above eigenvalues and condition (4.42) we can determine that the minimal commensurate order of this system is $q > 0.839$.

When we assume the commensurate order $q_1 = q_2 = q_3 = 0.9$ and parameters $a = 0.5$, $b = 0.2$ and $c = 10$, we get the following characteristic equation of the linearized system for the equilibrium E_1:

$$\lambda^{27} - 0.489989\lambda^{18} + 20.974974\lambda^{9} - 9.979979 = 0,$$

with unstable root $\lambda_1 \approx 0.9208$. The characteristic equation for the equilibrium E_2 is

$$\lambda^{27} + 9.489989\lambda^{18} - 3.974974\lambda^{9} + 9.979979 = 0,$$

with unstable roots $\lambda_{1,2} \approx 0.9892 \pm 0.1460j$, because $|\arg(\lambda_{1,2})| = 0.1466 < \pi/2m$, where $m = 10$ (LCM of orders denominator).

In Fig. 5.44 is depicted phase trajectory of the fractional-order Rössler's system (5.75) for commensurate order $q = 0.9$ and parameters $a = 0.5$, $b = 0.2$, $c = 10$, with the initial conditions $(x(0), y(0), z(0)) = (0.5, 1.5, 0.1)$, for simulation time $120s$ and time step $h = 0.005$.

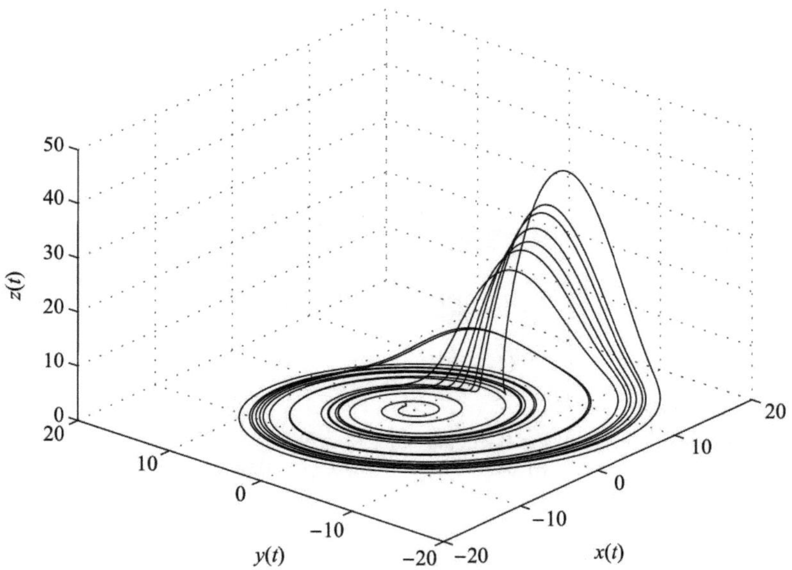

Fig. 5.44 Simulation result of the fractional-order Rössler's system (5.75) in state space for parameters $a = 0.5$, $b = 0.2$, $c = 10$ and orders $q_1 = q_2 = q_3 = 0.9$ for simulation time $120s$, with initial conditions $(x(0), y(0), z(0)) = (0.5, 1.5, 0.1)$.

When we assume the incommensurate order $q_1 = 0.90$, $q_2 = 0.85$, $q_3 = 0.95$ and parameters $a = 0.5$, $b = 0.2$ and $c = 10$, we get the following characteristic equation of the linearized system for the equilibrium E_1:

$$\lambda^{270} - 1/2\lambda^{185} + 0.010010\lambda^{175} + \lambda^{95} - 0.005005\lambda^{90} + 19.97997\lambda^{85} - 9.97997 = 0,$$

with unstable root $\lambda_1 \approx 0.9913$. The characteristic equation for the equilibrium E_2 is

$$\lambda^{270} - 1/2\lambda^{185} + 9.98998\lambda^{175} + \lambda^{95} - 4.994994\lambda^{90} + 9.97997 + 0.020020\lambda^{85} = 0,$$

with unstable roots $\lambda_{1,2} \approx 1.0000 \pm 0.0151j$, because $|\arg(\lambda_{1,2})| = 0.0151 < \pi/2m$, where $m = 100$ (LCM of orders denominator).

In Fig. 5.45 is depicted phase trajectory of the fractional-order Rössler's system (5.75) for incommensurate orders $q_1 = 0.90$, $q_2 = 0.85$, $q_3 = 0.95$ and parameters $a = 0.5$, $b = 0.2$, $c = 10$, with the initial conditions $(x(0), y(0), z(0)) = (0.5, 1.5, 0.1)$, for simulation time $120\,s$ and time step $h = 0.005$.

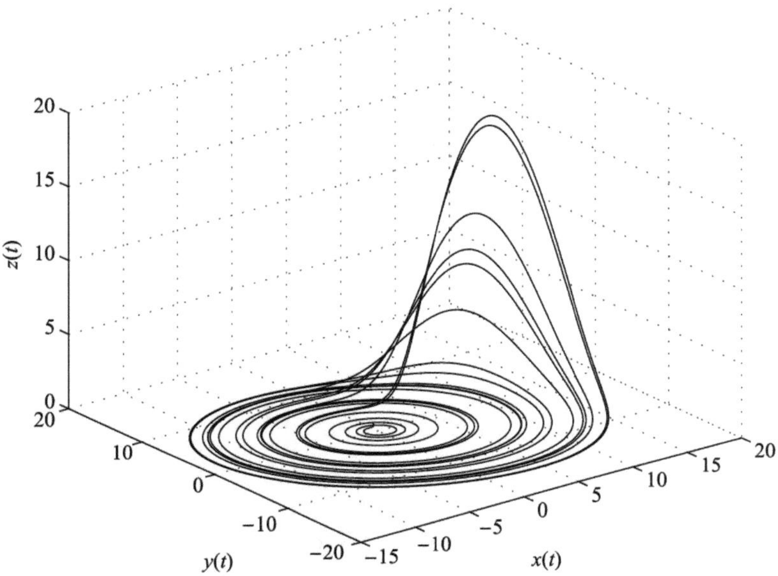

Fig. 5.45 Simulation result of the fractional-order Rössler's system (5.75) in state space for parameters $a = 0.5$, $b = 0.2$, $c = 10$ and orders $q_1 = 0.90$, $q_2 = 0.85$, $q_3 = 0.95$ for simulation time $120\,s$, with initial conditions $(x(0), y(0), z(0)) = (0.5, 1.5, 0.1)$.

The fractional-order Rössler hyperchaos equations were investigated in (Li and Chen, 2004), where chaotic behavior was cross-validated with the largest Lyapunov exponent.

5.12 Fractional-Order Newton-Leipnik's System

The Newton-Leipnik's system is described by the following nonlinear differential equations (Leipnik and Newton, 1981):

$$\begin{aligned}
\frac{dx(t)}{dt} &= -ax(t) + y(t) + 10y(t)z(t), \\
\frac{dy(t)}{dt} &= -x(t) - 0.4y(t) + 5x(t)z(t), \\
\frac{dz(t)}{dt} &= bz(t) - 5x(t)y(t),
\end{aligned} \tag{5.77}$$

where a and b are positive parameters.

In the paper (Leipnik and Newton, 1981), it has been noted that the Newton-Leipnik's system is a chaotic system with two strange attractors. When $a = 0.4$ and $b = 0.175$, with initial states $(0.349, 0, -0.16)$ and $(0.349, 0, -0.18)$, system (5.77) displays two strange attractors.

The system (5.77) with parameters $a = 0.4$ and $b = 0.175$ has five equilibrium points, where one of them is the origin. The other equilibria, approximately, are: $E_2 = (-0.23896; -0.03080; 0.21031)$, $E_3 = (-0.03154; 0.12237; -0.11031)$, $E_4 = (0.03154; -0.12237; -0.11031)$, and $E_5 = (0.23896; 0.03080; 0.21031)$.

The Jacobian matrix of the system (5.77) for equilibrium $E^* = (x^*, y^*, z^*)$ is

$$\mathbf{J} = \begin{bmatrix} -a & 1+10z^* & 10y^* \\ -1+5z^* & -0.4 & 5x^* \\ -5y^* & -5x^* & b \end{bmatrix}. \tag{5.78}$$

The eigenvalues of the Jacobian matrix (5.78) evaluated at all equilibrium points show that all equilibria are the saddle-focus points. For the equilibrium E_1 we obtain $\lambda_1 \approx 0.175$ and $\lambda_{2,3} \approx -0.4 \pm 1.0j$, for the equilibria E_2 and E_5 we get $\lambda_1 \approx -0.8$ and $\lambda_{2,3} \approx 0.0875 \pm 1.2113j$ and for the equilibria E_3 and E_4 we have $\lambda_1 \approx -0.8$ and $\lambda_{2,3} \approx 0.0875 \pm 0.8752j$. All these eigenvalues satisfy the condition for the system to be chaotic.

Here, the fractional-order Newton-Leipnik's system is considered, where integer-order derivative is replaced by a fractional one, as follows (Sheu et al., 2008):

$$\begin{aligned} {}_0D_t^{q_1} x(t) &= -ax(t) + y(t) + 10y(t)z(t), \\ {}_0D_t^{q_2} y(t) &= -x(t) - 0.4y(t) + 5x(t)z(t), \\ {}_0D_t^{q_3} z(t) &= bz(t) - 5x(t)y(t), \end{aligned} \tag{5.79}$$

where $0 < q_1, q_2, q_3 \leq 1$ are derivatives orders.

In the case of commensurate-order system, where $q_1 = q_2 = q_3 \equiv q$ we can determine a minimal order to satisfy a necessary condition (4.42) for chaotic behavior. For the equilibria E_2 and E_5 it is $q > 0.9540$ and for the equilibria E_3 and E_4 it is $q > 0.9365$.

Numerical solution of the fractional-order Newton-Leipnik's system (5.75) is given as follows:

$$\begin{aligned} x(t_k) &= \left(ax(t_{k-1}) + y(t_{k-1}) + 10y(t_{k-1})z(t_{k-1})\right) h^{q_1} - \sum_{j=v}^{k} c_j^{(q_1)} x(t_{k-j}), \\ y(t_k) &= \left(-x(t_k) - 0.4y(t_{k-1}) + 5x(t_k)z(t_{k-1})\right) h^{q_2} - \sum_{j=v}^{k} c_j^{(q_2)} y(t_{k-j}), \\ z(t_k) &= \left(bz(t_{k-1}) - 5x(t_k)y(t_k)\right) h^{q_3} - \sum_{j=v}^{k} c_j^{(q_3)} z(t_{k-j}), \end{aligned} \tag{5.80}$$

where T_{sim} is the simulation time, $k = 1,2,3,\ldots,N$, for $N = [T_{sim}/h]$, and $(x(0)$, $y(0)$, $z(0))$ is the start point (initial conditions). The binomial coefficients $c_j^{(q_i)}$, $\forall i$ are calculated according to the relation (2.54).

Let us consider the following parameters $a = 0.4$, $b = 0.175$, orders $q_1 = 0.95$, $q_2 = 0.95$, and $q_3 = 0.95$ of the system (5.79).

In Fig. 5.46 is depicted phase trajectory for derivative orders $q_1 = 0.95$, $q_2 = 0.95$, $q_3 = 0.95$, parameters $a = 0.4$, $b = 0.175$ for simulation time $200\,s$, time step $h = 0.005$ and with the initial conditions: $(x(0), y(0), z(0)) = (0.19, 0.0, -0.18)$. In the paper (Sheu et al., 2008). it has been noted that the system still approaches the same attractor for both initial states $(0.349, 0, -0.16)$ and $(0.349, 0, -0.18)$, and we can confirm it. Because of this we use a different set of initial conditions.

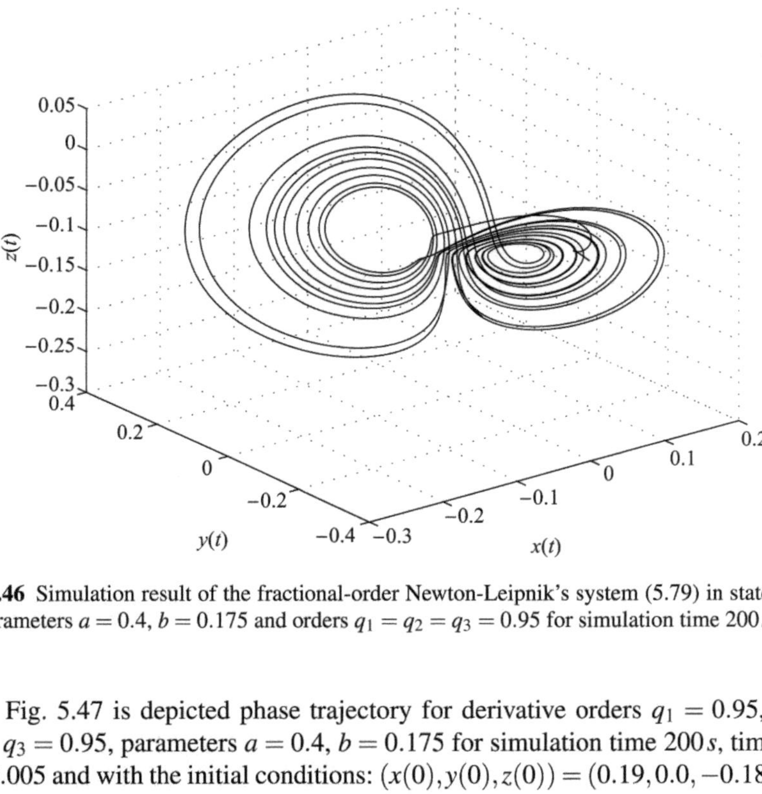

Fig. 5.46 Simulation result of the fractional-order Newton-Leipnik's system (5.79) in state space for parameters $a = 0.4$, $b = 0.175$ and orders $q_1 = q_2 = q_3 = 0.95$ for simulation time $200\,s$.

In Fig. 5.47 is depicted phase trajectory for derivative orders $q_1 = 0.95$, $q_2 = 0.95$, $q_3 = 0.95$, parameters $a = 0.4$, $b = 0.175$ for simulation time $200\,s$, time step $h = 0.005$ and with the initial conditions: $(x(0), y(0), z(0)) = (0.19, 0.0, -0.18)$ projected onto $x - y$ plane. We can observe that double scroll attractor surrounded the equilibria E_3 and E_4.

The characteristic equation of the linearized system evaluated at the equilibrium E_3 or E_4 is

$$\lambda^{285} + 5/8\lambda^{190} + 0.63369\lambda^{95} + 0.618953 = 0,$$

with unstable roots $\lambda_{1,2} \approx 0.9985 \pm 0.0155j$, because $|\arg(\lambda_{1,2})| = 0.0155 < \pi/2m$, where $m = 100$ (LCM of orders denominator).

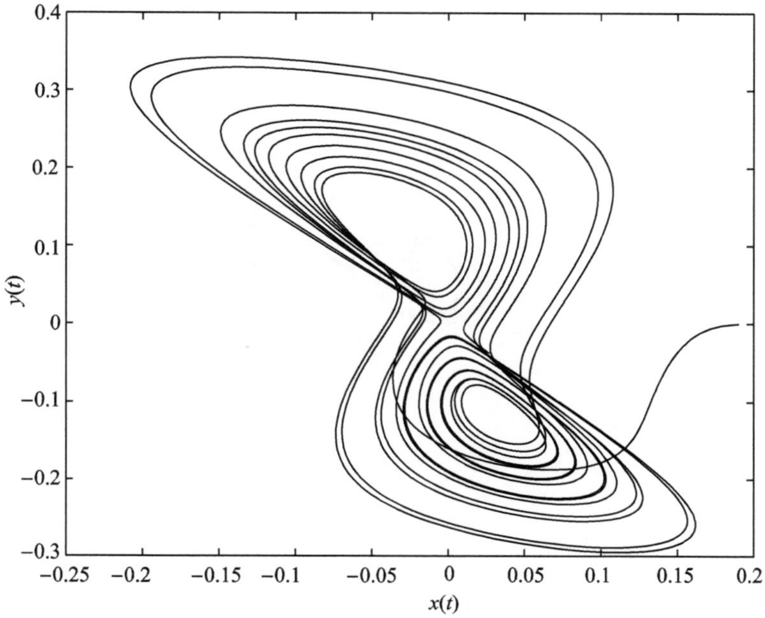

Fig. 5.47 Simulation result of the fractional-order Newton-Leipnik's system (5.79) projected onto $x-y$ plane for parameters $a = 0.4$, $b = 0.175$ and orders $q_1 = q_2 = q_3 = 0.95$.

The characteristic equation of the linearized system evaluated at the equilibrium E_1 is

$$\lambda^{285} + 5/8\lambda^{190} + 1.020\lambda^{95} - 0.2030 = 0$$

with unstable root $\lambda \approx 0.9818$. The equilibria E_2 and E_5 are stable for the above parameters and orders of the system.

Let us consider the following parameters $a = 0.4$, $b = 0.175$, orders $q_1 = 1$, $q_2 = 0.97$, and $q_3 = 1$ of the system (5.79).

In Fig. 5.48 is depicted phase trajectory for derivative orders $q_1 = q_3 = 1$, $q_2 = 0.97$, parameters $a = 0.4$, $b = 0.175$ for simulation time $200\,s$, time step $h = 0.005$ and with the initial conditions: $(x(0), y(0), z(0)) = (-0.8, 0.0, 0.18)$.

In Fig. 5.49 is depicted phase trajectory for derivative orders $q_1 = q_3 = 1$, $q_2 = 0.97$, parameters $a = 0.4$, $b = 0.175$ for simulation time $200\,s$, time step $h = 0.005$ and with the initial conditions: $(x(0), y(0), z(0)) = (-0.8, 0.0, 0.18)$ projected onto $x-y$ plane. We can observe that double scroll attractor surrounded the equilibria E_2 and E_5.

The characteristic equation of the linearized system evaluated at the equilibrium E_2 or E_5 is

$$\lambda^{297} + 2/5\lambda^{200} + 9/40\lambda^{197} + 1.35761\lambda^{100} - 0.02255\lambda^{97} + 1.18004 = 0,$$

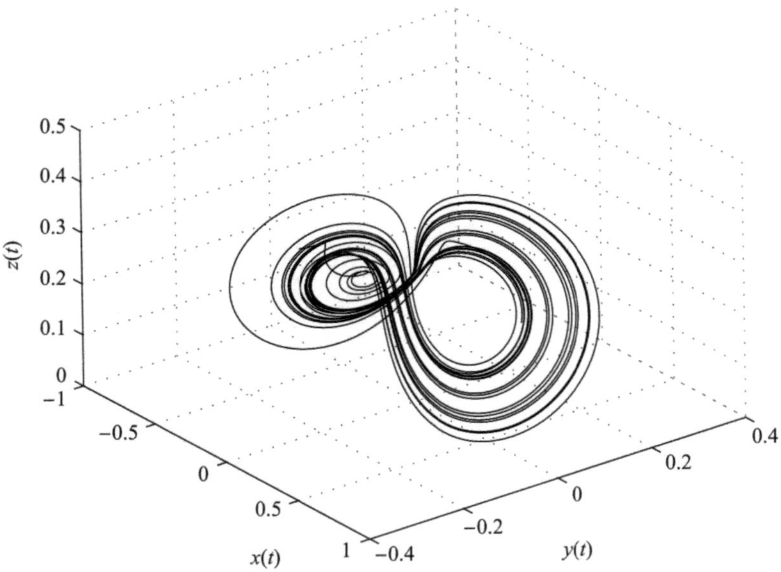

Fig. 5.48 Simulation result of the fractional-order Newton-Leipnik's system (5.79) in state space for parameters $a = 0.4$, $b = 0.175$ and orders $q_1 = q_3 = 1$, $q_2 = 0.97$ for simulation time $200\,s$.

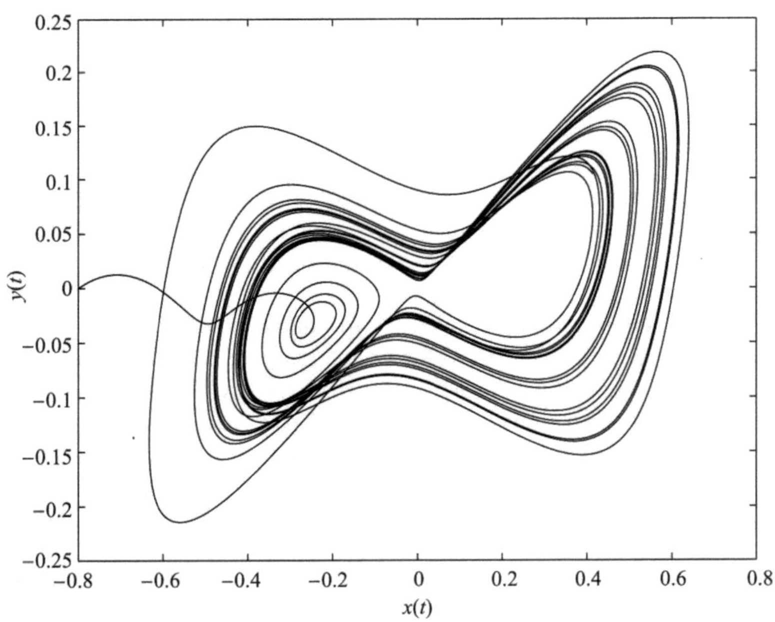

Fig. 5.49 Simulation result of the fractional-order Newton-Leipnik's system (5.79) projected onto $x - y$ plane for parameters $a = 0.4$, $b = 0.175$ and orders $q_1 = q_3 = 1$, $q_2 = 0.97$.

with unstable roots $\lambda_{1,2} \approx 1.0018 \pm 0.0152j$, because $|\arg(\lambda_{1,2})| = 0.0152 < \pi/2m$, where $m = 100$.

The characteristic equation of the linearized system evaluated at the equilibrium E_1 is

$$\lambda^{297} + 9/40\lambda^{197} + 2/5\lambda^{200} + 1.090\lambda^{100} - 7/100\lambda^{97} - 0.203 = 0,$$

with unstable root $\lambda \approx 0.9827$.

The characteristic equation of the linearized system evaluated at the equilibrium E_3 or E_4 is

$$\lambda^{297} + 2/5\lambda^{200} + 9/40\lambda^{197} - 0.04511\lambda^{100} + 0.67880\lambda^{97} + 0.61895 = 0,$$

with unstable roots $\lambda_{1,2} \approx 0.9986 \pm 0.0148j$, because $|\arg(\lambda_{1,2})| = 0.0148 < \pi/2m$, where $m = 100$.

It is interesting to observe that the system has two strange attractors to rotate by approximately 90° as depicted in Fig. 5.50. The dynamic of the fractional-order Newton-Leipnik's system was studied (Sheu et al., 2008), where the lowest total order of the system to yield chaos was found to be 2.82. This system displays better dynamic behavior.

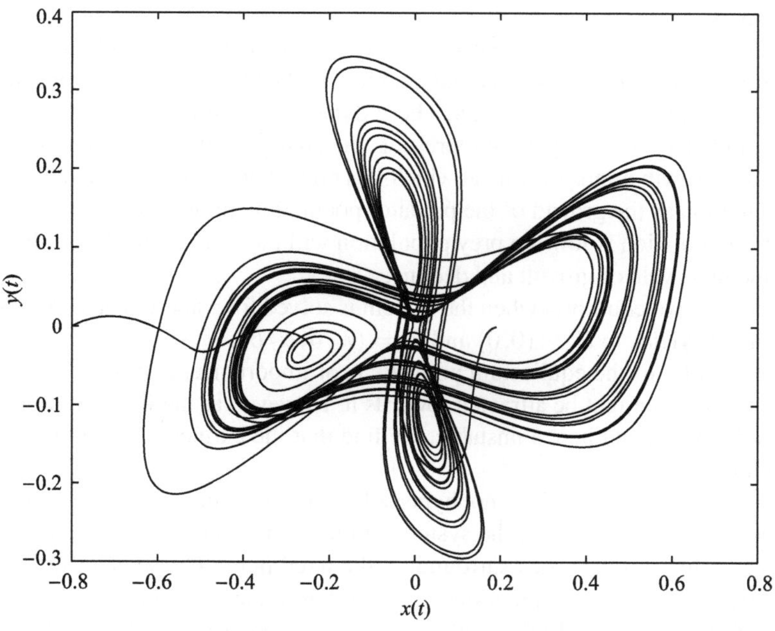

Fig. 5.50 Simulation results of the fractional-order Newton-Leipnik's system (5.79) projected onto $x-y$ plane for parameters $a = 0.4$, $b = 0.175$, orders $q_1 = q_2 = q_3 = 0.95$ and $q_1 = q_3 = 1$, $q_2 = 0.97$, respectively, for simulation time 200 s.

5.13 Fractional-Order Lotka-Volterra System

The Lotka-Volterra equations, also known as the predator-prey (or parasite-host) equations, are a pair of first order, non-linear, differential equations frequently used to describe the dynamics of biological systems in which two species interact on each other, one is a predator and the other is its prey. They were proposed independently by Alfred J. Lotka in 1925 and Vito Volterra in 1926.

Classical integer-order model of the Lotka-Volterra system is deffined as

$$\begin{aligned} \frac{dx(t)}{dt} &= x(t)(\alpha - \beta y(t)) \\ \frac{dy(t)}{dt} &= -y(t)(\gamma - \delta x(t)), \end{aligned} \tag{5.81}$$

where $y(t)$ is the number of some predators (for example, wolves); $x(t)$ is the number of its prey (for example, rabbits); $dy(t)/dt$ and $dx(t)/dt$ represent the growth of the two populations against time; t represents the time; and α, β, γ and δ are parameters representing the interaction of the two species.

The equations have periodic solutions which do not have a simple expression in terms of the usual trigonometric functions. However, an approximate linearized solution yields a simple harmonic motion with the population of predators following that of prey by 90°.

In the model system, the predators thrive when there are plentiful prey but, ultimately, outstrip their food supply and decline. As the predator population is low the prey population will increase again. These dynamics continue in a cycle of growth and decline. Hence the equation represents the change in the prey's numbers given by its own growth minus the rate at which it is preyed upon; the change in the predator population as the growth of the predator population, minus natural death. As the predator population is low the prey population will increase again. These dynamics continue in a cycle of growth and decline.

There are two equilibria when the system is solved for x and y. The above system of equations yields to $E_1 = (0;0)$ and $E_2 = (\lambda/\delta;\alpha/\beta)$.

The stability of the equlibrium point E_1 is of importance. If it were stable, non-zero populations might be attracted towards it. However, as the fixed point in origin is a saddle point, and hence unstable, we find that the extinction of both species is difficult in the model.

The second fixed point E_2 is not hyperbolic, so no conclusions can be drawn from the linear analysis. However, the system admits a constant of motion and the level curves are closed trajectories surrounding the fixed point. Consequently, the levels of the predator and prey populations cycle, and oscillate around this fixed point.

The fractional-order Lotka-Volterra (or fractional-order predator-prey model) system was proposed and described as (Ahmed et al., 2007):

$$\begin{aligned} {}_0D_t^q x(t) &= x(t)(\alpha - rx(t) - \beta y(t)) \\ {}_0D_t^q y(t) &= -y(t)(\gamma - \delta x(t)), \end{aligned} \tag{5.82}$$

where $0 < q \leqslant 1$, $x \geqslant 0$, $y \geqslant 0$ are prey and predator densities, respectively, and all constants r, α, β, γ and δ are positive. For $r = 0$ and $q = 1$ we obtain a well-known model (5.81). The stability analysis and numerical solutions of such kind of system have been studied (Ahmed et al., 2007).

In the paper (Samardzija and Greller, 1988) was proposed a two-predator and one-prey generalization of the Lotka-Volterra system. We assume its fractional-order modification as follows:

$$\begin{aligned}
{}_0D_t^{q_1}x(t) &= ax(t) - bx(t)y(t) + ex^2(t) - sz(t)x^2(t),\\
{}_0D_t^{q_2}y(t) &= -cy(t) + dx(t)y(t),\\
{}_0D_t^{q_3}z(t) &= -pz(t) + sz(t)x^2(t),
\end{aligned} \tag{5.83}$$

where a, b, c, d, e, p, s are model parameters and q_1, q_2, q_3 are fractional orders. When we consider $p = 0$, $s = 0$, $q_1 = q_2$ and $e = -r$ in the general model (5.83), we obtain the fractional-order (one predator and one prey) Lotka-Volterra model (5.82) upon the substitutions $\alpha \equiv a$, $\beta \equiv b$, $\gamma \equiv c$, and $\delta \equiv d$.

The proposed fractional-order Lotka-Volterra system (5.83) has five equilibrium points: $E_1 = (0;0;0)$, $E_2 = (-a/e;0;0)$, $E_3 = (\sqrt{sp}/s;0;(a + (e\sqrt{sp})/s)/\sqrt{sp})$, $E_4 = (-\sqrt{sp}/s;0;-(a - (e\sqrt{sp})/s)/\sqrt{sp})$, and $E_5 = (c/d;(da + ec)/db;0)$;

The Jacobian matrix of the system (5.83) for equilibrium $E^* = (x^*, y^*, z^*)$ is

$$\mathbf{J} = \begin{bmatrix} a - by^* + 2ex^* - 2sz^*x^* & -bx^* & -sx^{*2} \\ dy^* & -c + dx^* & 0 \\ 2sx^*z^* & 0 & -p + sx^{*2} \end{bmatrix}. \tag{5.84}$$

Numerical solution of the fractional-order Lotka-Volterra system (5.83) is given as follows:

$$\begin{aligned}
x(t_k) &= \left(x(t_{k-1})(a - by(t_{k-1}) + ex(t_{k-1}) - sz(t_{k-1})x(t_{k-1}))\right)h^{q_1} - \sum_{j=v}^{k} c_j^{(q_1)} x(t_{k-j}),\\
y(t_k) &= \left(-cy(t_{k-1}) + dx(t_k)y(t_{k-1})\right)h^{q_2} - \sum_{j=v}^{k} c_j^{(q_2)} y(t_{k-j}),\\
z(t_k) &= \left(-pz(t_{k-1}) + sz(t_{k-1})x^2(t_k)\right)h^{q_3} - \sum_{j=v}^{k} c_j^{(q_3)} z(t_{k-j}),
\end{aligned} \tag{5.85}$$

where T_{sim} is the simulation time, $k = 1, 2, 3, \ldots, N$, for $N = [T_{sim}/h]$, and ($x(0)$, $y(0)$, $z(0)$) is the start point (initial conditions). The binomial coefficients $c_j^{(q_i)}$, $\forall i$ are calculated according to the relation (2.54).

Let us consider the following system parameters $a = 2, b = 1, c = 3, d = 1, e = 0, p = 0, s = 0$ and derivative orders $q_1 = q_2 = q_3 = 1$ and $q_1 = q_2 = q_3 = 0.9$, respectively. For these parameters the system (5.83) has two equilibria $E_1 = (0;0)$ and $E_2 = (3;2)$ and their corresponding eigenvalues are $\lambda_1 = 2$ and $\lambda_2 = -3$ for E_1

and $\lambda_{1,2} = \pm 2.4495j$ for E_2. Equilibrium E_1 is a saddle point and equilibrium E_2 is a center. The fixed point E_2 is not hyperbolic.

In Fig. 5.51 and Fig. 5.52 are depicted phase trajectories for various derivative orders $\bar{q} = 1.0$ and $\bar{q} = 0.9$, respectively, for simulation time $60\,s$, time step $h = 0.005$ and for the initial conditions: $(x(0), y(0), z(0)) = (1, 2, 0)$.

Let us consider the following system parameters $a = 1, b = 1, c = 1, d = 1, e = 2, p = 3, s = 2.7$ and orders $q_1 = q_2 = q_3 = 0.95$.

In Fig. 5.53 and Fig. 5.54 are depicted phase trajectories of the fractional-order Lotka-Volterra system (5.83) for orders $q_1 = q_2 = q_3 = 0.95$, and parameter $a = 1, b = 1, c = 1, d = 1, e = 2, p = 3, s = 2.7$, for simulation time $200\,s$, time step $h = 0.005$, and for the initial conditions: $(x(0), y(0), z(0)) = (1, 1.4, 1)$. We can observe the so-called "fractal torus" (Samardzija and Greller, 1988).

For the above parameters we obtain the following values of equilibrium points $E_1 = (0;0;0), E_2 = (-0.5;0;0), E_3 = (-1.0540;0;0.3893), E_4 = (1.0540;0;1.0921)$, and $E_5 = (1;3;0)$. The corresponding eigenvalues of the Jacobian matrix (5.84) evaluated at equilibrium points are: $\lambda_1 = -3$, $\lambda_2 = -1$, $\lambda_3 = 1$ for E_1, $\lambda_1 - -1$, $\lambda_2 = -1.5$, $\lambda_3 = -2.325$ for E_2, $\lambda_1 \approx -3.1266$, $\lambda_2 \approx 2.12661$, $\lambda_3 \approx -2.05409$ for E_3, $\lambda_1 = 0.5409$, $\lambda_{2,3} \approx -0.50 \pm 4.2894j$ for E_4, and $\lambda_1 = -0.3$, $\lambda_{2,3} \approx 1.0 \pm 1.4142j$ for E_5. The equlibrium E_1 is saddle point, the equilibrium E_2 is stable node, the equlibrium E_3 is saddle point, the equilibria E_4 and E_5 are saddle-focus points.

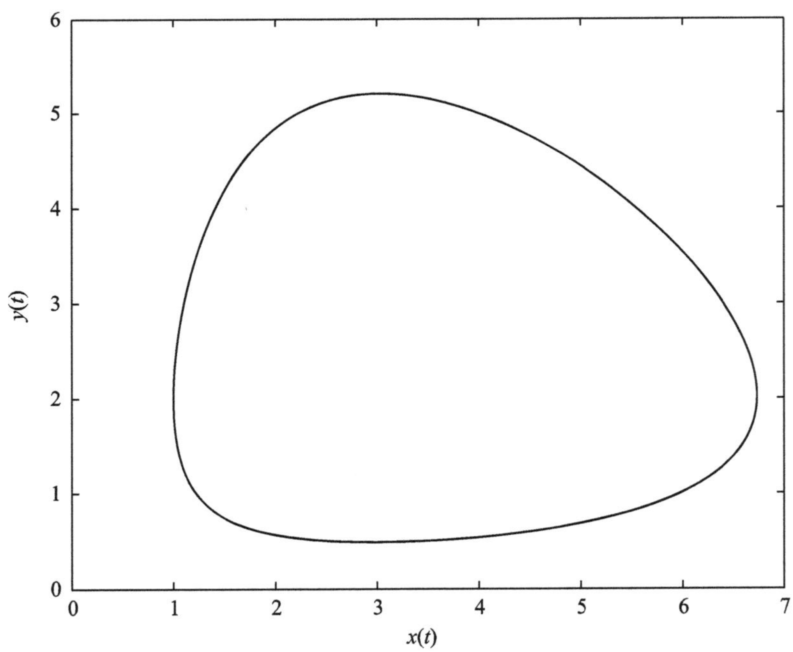

Fig. 5.51 Phase plane $x - y$ trajectory (limit cycle) for the Lotka-Volterra system with orders $q_1 = q_2 = 1.0$, $q_3 = 0$, and parameters $a = 2, b = 1, c = 3, d = 1, e = 0, p = 0, s = 0$.

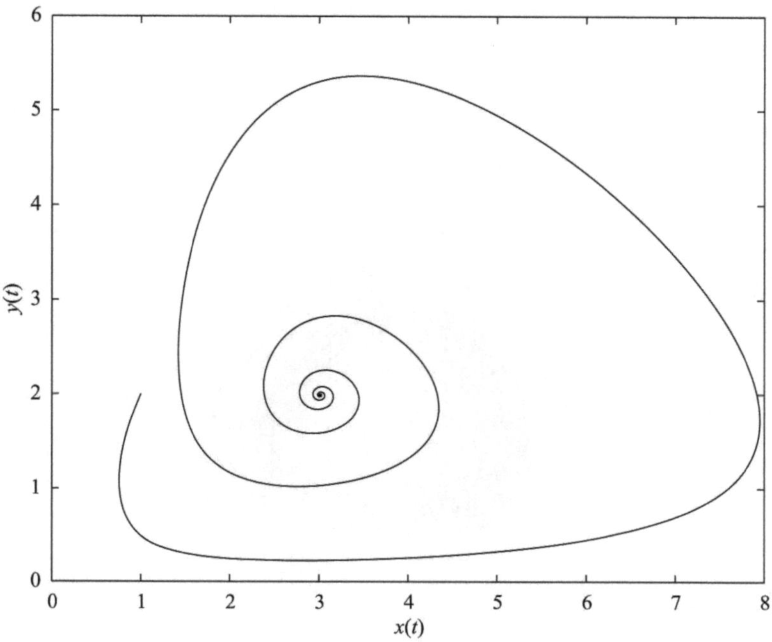

Fig. 5.52 Phase plane $x-y$ trajectory for the Lotka-Volterra with fractional-orders $q_1 = q_2 = q_3 = 0.9$, and parameters $a = 2, b = 1, c = 3, d = 1, e = 0, p = 0, s = 0$.

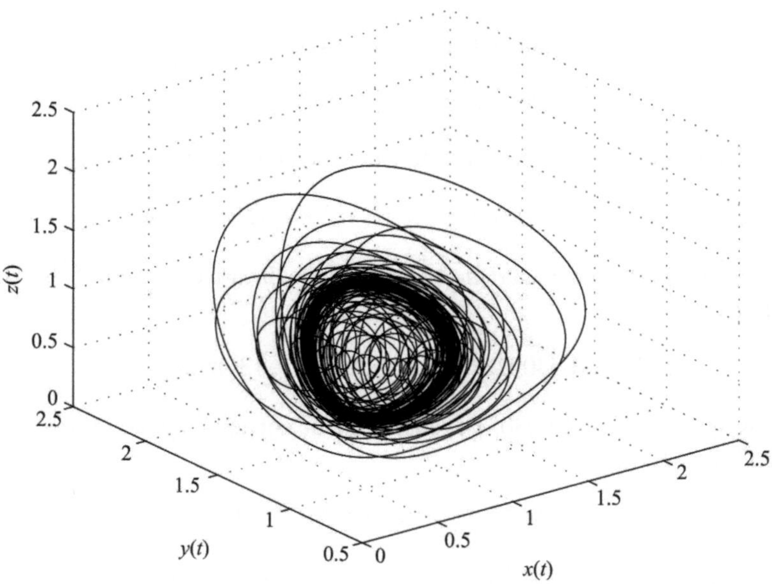

Fig. 5.53 Phase trajectory of the Lotka-Volterra system with orders $q_1 = q_2 = q_3 = 0.95$ and parameters $a = 1, b = 1, c = 1, d = 1, e = 2, p = 3, s = 2.7$ in state space.

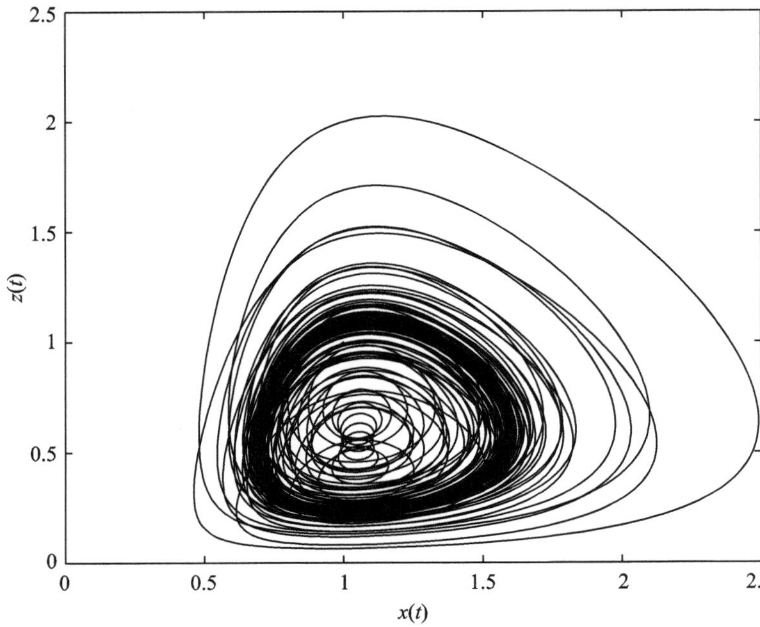

Fig. 5.54 Phase trajectory for the Lotka-Volterra with fractional-orders $q_1 = q_2 = q_3 = 0.95$, and parameters $a = 1, b = 1, c = 1, d = 1, e = 2, p = 3, s = 2.7$ in state plane $x - z$.

The characteristic equation of the linearized system (5.83) at the equlibrium point E_1 is

$$(\lambda^{95} - 1)(\lambda^{95} + 1)(\lambda^{95} + 3) = 0,$$

with unstable root $\lambda_1 = 1$. The characteristic equation at equilibrium E_2 is

$$(\lambda^{95} + 1.0)(\lambda^{95} + 1.5)(\lambda^{95} + 2.3250) = 0,$$

with stable roots. The characteristic equation of the linearized system (5.83) at equilibrium E_3 is

$$\lambda^{285} + 3.054092\lambda^{190} - 4.59501\lambda^{95} - 13.657888 = 0,$$

with unstable root $\lambda_1 \approx 1.0080$. The characteristic equation of the linearized system at equilibrium E_4 is

$$\lambda^{285} + 0.945907\lambda^{190} + 18.595018\lambda^{95} - 1.008778 = 0,$$

with unstable root $\lambda_1 \approx 0.9698$. Finally, the characteristic equation at equilibrium point E_5 is

$$\lambda^{285} - 1.7\lambda^{190} + 2.4\lambda^{95} + 0.9 = 0,$$

with unstable roots $\lambda_{1,2} \approx 1.0057 \pm 0.0101j$, because $|\arg(\lambda_{1,2})| = 0.0101 < \pi/2m$, where $m = 100$ (LCM of orders denominator). The condition for the chaotic system is satisfied.

The "fractal torus" attractor exhibited by the system (5.83) for certain values of the parameters and orders is interesting because it exhibits a structure entirely different from attractors such as, for instance, the Rössler or Lorenz attractors. Numerical simulations showed that different initial conditions often lead to different fast manifolds.

The system described in this section suggests that it is a reasonable biological or chemical model. It could be modified also to one-predator and two-prey scheme.

5.14 Fractional-Order Financial System

The chaotic phenomenon in macro economics was found in 1985. The continuos economical system was described and analyzed (Ma and Chen, 2001a,b). The simplified financial model is defined as:

$$\begin{aligned} \frac{dx(t)}{dt} &= z(t) + (y(t) - a)x(t), \\ \frac{dy(t)}{dt} &= 1 - by(t) - x(t)^2, \\ \frac{dz(t)}{dt} &= -x(t) - cz(t), \end{aligned} \tag{5.86}$$

where a is the saving amount, b is the cost per investment, and c is the elasticity of demand of commercial market. The state variables are: $x(t)$ is the interest rate, $y(t)$ is the investment demand, and $z(t)$ is the price index.

The system (5.86) has three equilibrium points: $E_1 = (0; 1/b; 0)$,

$$E_2 = (\sqrt{(c-b-abc)/c}; (1+ac)/c; -(1/c)\sqrt{(c-b-abc)/c},$$

$$E_3 = (-\sqrt{(c-b-abc)/c}; (1+ac)/c; (1/c)\sqrt{(c-b-abc)/c}.$$

The Jacobian matrix of the system (5.86), evaluated at the equilibrium $E^* = (x^*, y^*, z^*)$, is given by

$$\mathbf{J} = \begin{bmatrix} -a+y^* & x^* & 1 \\ -2^* & -b & 0 \\ -1 & 0 & -c \end{bmatrix}. \tag{5.87}$$

The fractional-order financial system is described as follows (Chen, 2008):

$$
\begin{aligned}
{}_0D_t^{q_1}x(t) &= z(t)+(y(t)-a)x(t),\\
{}_0D_t^{q_2}y(t) &= 1-by(t)-x(t)^2,\\
{}_0D_t^{q_3}z(t) &= -x(t)-cz(t),
\end{aligned}
\tag{5.88}
$$

where the total order of the system is denoted by $\bar{q} = (q_1, q_2, q_3)$.

The numerical solution of the fractional-order financial system has the following form:

$$
\begin{aligned}
x(t_k) &= (z(t_{k-1}) - (y(t_{k-1}) - a)x(t_{k-1}))\, h^{q_1} - \sum_{j=v}^{k} c_j^{(q_1)} x(t_{k-j}),\\
y(t_k) &= \left(1 - by(t_k) - x^2(t_k)\right) h^{q_2} - \sum_{j=v}^{k} c_j^{(q_2)} y(t_{k-j}),\\
z(t_k) &= (-x(t_k) - cz(t_{k-1}))\, h^{q_3} - \sum_{j=v}^{k} c_j^{(q_3)} z(t_{k-j}),
\end{aligned}
\tag{5.89}
$$

where T_{sim} is the simulation time, $k = 1,2,3,\ldots,N$, for $N = [T_{sim}/h]$, and ($x(0)$, $y(0)$, $z(0)$) is the start point (initial conditions). The binomial coefficients $c_j^{(q_i)}$, $\forall i$ are calculated according to the relation (2.54).

Let us consider the following parameters of the system (5.88): $a = 1.0$, $b = 0.1$ and $c = 1.0$. The corresponsing eigenvalues for the system equilibrium $E_1 = (0;10;0)$ are: $\lambda_1 \approx 8.898979$, $\lambda_2 \approx -0.8989794$, and $\lambda_3 \approx -0.1$. It is a saddle point. For the equilibrium points $E_2 = (0.894427;2;-0.894427)$ and $E_3 = (-0.894427;2;0.894427)$ they are: $\lambda_1 \approx -0.7608747$ and $\lambda_{2,3} \approx 0.3304373 \pm 1.411968j$. It is a saddle-focus point. Because it is an unstable equilibrium, the condition for chaos is satisfied and chaotic system (5.88) with the above parameters can exhibit chaotic behavior. The minimal commensurate order of the system is $q > 0.8536$.

Assume the commensurate order $q_1 = q_2 = q_3 = 0.9$ of the system (5.88) with the parameters $a = 1.0$, $b = 0.1$, and $c = 1.0$. The characteristic equation of the linearized system for equilibrium E_1 is

$$\lambda^{27} - 7.9\lambda^{18} - 8.8\lambda^{9} - 0.8 = 0$$

and the unstable root for the equilibrium E_1 is $\lambda_1 \approx 1.2749$. The characteristic equation of the linearized system for equilibria E_2 and E_3 is

$$\lambda^{27} + 0.1\lambda^{18} + 1.6\lambda^{9} + 1.6 = 0$$

and unstable eigenvalues for equilibria E_2 and E_3 are $\lambda_{1,2} \approx 1.0306 \pm 0.1547j$, because $|\arg(\lambda_{1,2})| = 0.1490 < \pi/2m$, where $m = 10$ (LCM of orders denominator).

In Fig. 5.55 is depicted the simulation result of the financial system (5.88) for the following parameters: $a = 1$, $b = 0.1$, and $c = 1.0$, orders $q_1 = 0.9$, $q_2 = 0.9$, $q_3 = 0.9$, and computational time $200\,s$, for time step $h = 0.005$.

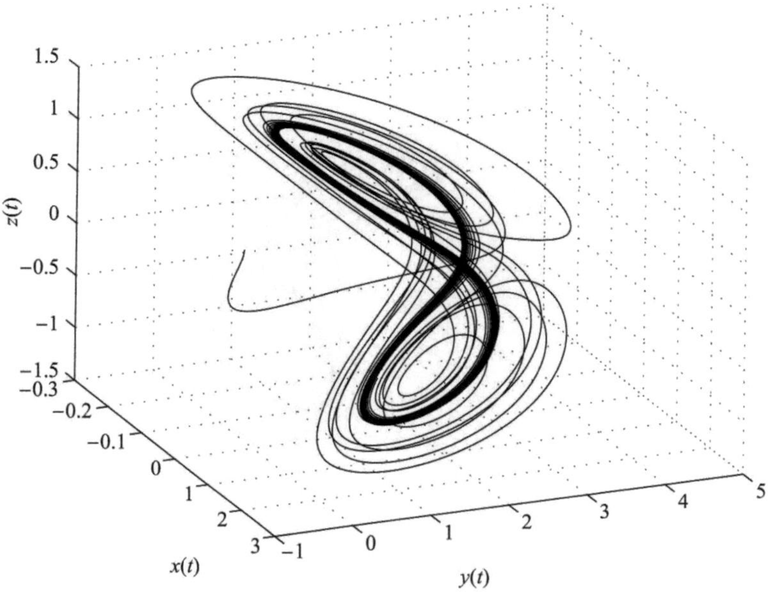

Fig. 5.55 Simulation result of the fractional-order financial system (5.88) in state space for the initial conditions $(x(0), y(0), z(0)) = (2, -1, 1)$.

Assume the incommensurate order $q_1 = 1.0$, $q_2 = 0.95$, $q_3 = 0.9$ of the system (5.88) with the parameters $a = 1.0$, $b = 0.1$, and $c = 1.0$. The characteristic equation of the linearized system for equilibrium E_1 is

$$\lambda^{285} + \lambda^{195} + 1/10\lambda^{190} - 9\lambda^{185} + 1/10\lambda^{100} - 8\lambda^{95} - 0.9\lambda^{90} - 0.8 = 0$$

and unstable root for the equilibrium E_1 is $\lambda_1 \approx 1.0221$. The characteristic equation of the linearized system for equilibria E_2 and E_3 is

$$\lambda^{285} + \lambda^{195} + 1/10\lambda^{190} - \lambda^{185} + 1/10\lambda^{100} + 1.5\lambda^{90} + 1.6 = 0$$

and unstable eigenvalues for equilibria E_2 and E_3 are $\lambda_{1,2} \approx 1.0035 \pm 0.0139j$, because $|\arg(\lambda_{1,2})| = 0.0138 < \pi/2m$, where $m = 100$ (LCM of orders denominator).

In Fig. 5.56 are depicted the simulation results of the financial system (5.88) for the following parameters: $a = 1.0$, $b = 0.1$, and $c = 1.0$, orders $q_1 = 1.0$, $q_2 = 0.95$, $q_3 = 0.9$, and computational time $200\,s$, for time step $h = 0.005$.

Investigation of chaos in various cases of the fractional-order financial system and its cross-validation with the largest Lyapunov exponent was done (Chen, 2008). The lowest order at which this system yielded chaos was 2.35.

In the next chapter we will use this system as a controlled system for the sliding mode control strategy.

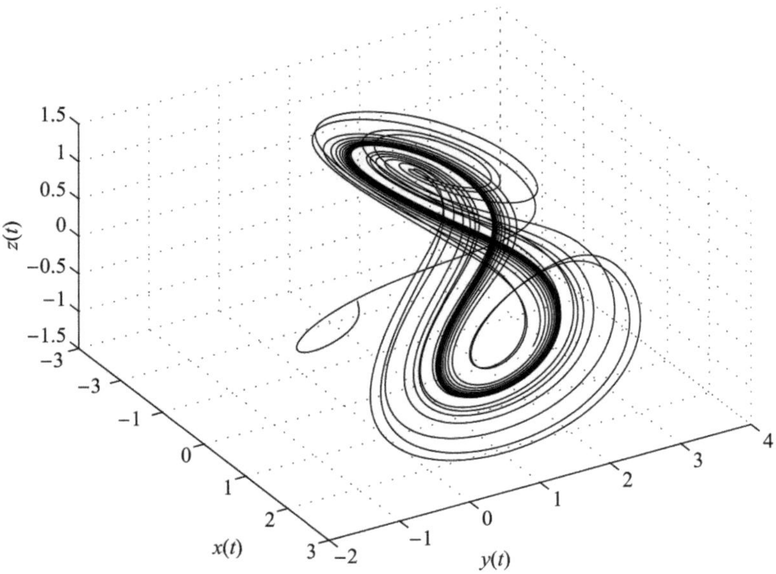

Fig. 5.56 Simulation result of the fractional-order financial system (5.88) in state space for the initial conditions $(x(0), y(0), z(0)) = (2, -1, 1)$.

5.15 Fractional-Order CNN

The basic circuit unit of the Cellular Neural Network (CNN) is a cell. The CNN was introduced by L. O. Chua in 1988. It contains linear and non-linear circuit elements, which typically are: linear capacitor, linear resistors, linear and non-linear controlled sources, and independent sources. Any cell in the CNN is connected only to its neighbor cells. Theoretically we can define the CNN of any dimension, e.g. two-dimensional array of $M \times N$, having M rows and N columns. We call the cell on the i-th row and the j-th column the cell $C(i, j)$. Observe from Fig. 5.57 that each cell $C(i, j)$ contains one independent voltage source $E_{i,j}$, one independent current source I, one linear capacitor C, two linear resistors R_x and R_y, controlling input voltage u_{ij}, state voltage of the cell x_{ij}, feedback from the output voltage y_{ij} of each neighbor cell $C(k, l)$. In fact each cell $C(i, j)$ mutually interacts with all the cells belonging to its neighbors $N_r(i, j)$ by means of the voltage-controlled current source $I_{xy}(i, j; k, l) = A(i, j; k, l)y_{kl}$, $I_{xu}(i, j; k, l) = B(i, j; k, l)u_{kl}$ and $I_{xx}(i, j; k, l) = C(i, j; k, l)x_{kl}$. The constant coefficients $A(i, j; k, l)$, $B(i, j; k, l)$ and $C(i, j; k, l)$ are known as the cloning templates. If they are equal for each cell, they are called space-invariant. The CNN is described by the following state equation of all its cells (Chua and Roska, 1993):

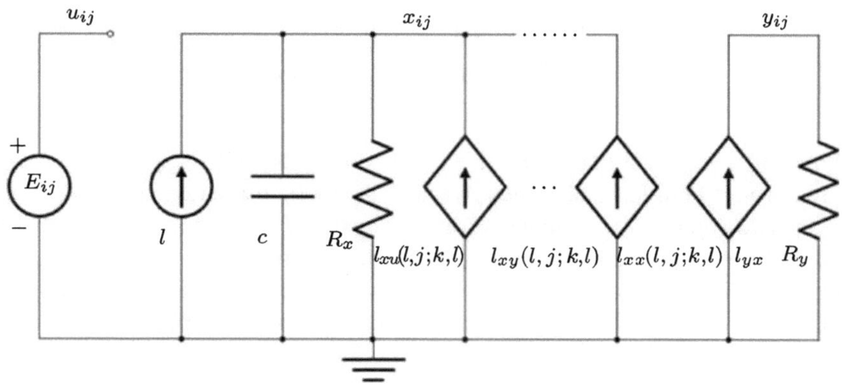

Fig. 5.57 The CNN cell

$$C\frac{dx_{ij}(t)}{dt} = -\frac{1}{R_x}x_{ij}(t) + \sum_{C(k,l)\in N_r(i,j)} A(i,j;k,l)y_{kl}(t) + \sum_{C(k,l)\in N_r(i,j)} B(i,j;k,l)u_{kl}(t) + C(i,j;k,l)x_{kl}(t) + I, \quad (5.90)$$

with $x_{ij}(0) = x_{ij0}$, $C > 0$, $R_x > 0$, $1 \le i \le M$, and $1 \le j \le N$, where

$$N_r(i,j) = \{C(k,l) : \max(|k-i| - |l-j|) \le r\}$$

is the r-neighborhood. Input equation is: $u_{ij}(t) = E_{ij}$. Output equation is:

$$f(x_{ij}) = y_{ij}(t) = \frac{1}{2}(|x_{ij}(t)+1| - |x_{ij}(t)-1|). \quad (5.91)$$

This model with direct dependence of state variable on the state of the neighboring cells is known as a state-controlled CNN. Such kind of CNNs is also able to show chaotic behavior (Chua and Roska, 1993; Biey et al., 2003; Zou and Nossek, 1993).

The only non-linear element in each cell is a piecewise-linear voltage-controlled current source: $I_{yx} = (1/R_y)f(x_{ij})$.

In addition, in this section we derive the fractional-order model of the CNN described by (5.90). For this purpose we will consider the general model of a capacitor described by Equation (2.75). Westerlund and Ekstam provided in their work (Westerlund and Ekstam, 1994) the table of various capacitor dielectric with appropriate constant m (derivative order), which has been obtained experimentally by measurements. Carlson also studied, in 1963, the fractional capacitor and appropriate approximation technique for its model (Carlson and Halijak, 1963).

Applying the Kirchhoff law and the relation (2.75) to standard model of the CNN which is described by Equation (5.90), we obtain a fractional-order model of the CNN in the following form (Petráš, 2006):

$$C\frac{d^m x_{ij}(t)}{dt^m} = -\frac{1}{R_x}x_{ij}(t) + \sum_{C(k,l)\in N_r(i,j)} A(i,j;k,l)y_{kl}(t) + \sum_{C(k,l)\in N_r(i,j)} B(i,j;k,l)u_{kl}(t) + C(i,j;k,l)x_{kl}(t) + I, \quad (5.92)$$

with $x_{ij}(0) = x_{ij0}$, $C > 0$, $0 < m < 1$, $R_x > 0$, $1 \leq i \leq M$, and $1 \leq j \leq N$.

In the works (Arena at al., 1998; Arena et al., 2000), the parameter m is $1 < m < 1.5$ and two-cell CNN was studied. Considering that $0 < m < 1$ and the fact that we would like to study the behavior of system with the total order less than three, we have to consider three-cell fractional-order CNN. Referring to the general definition of CNN given by (5.92) and choosing the opposite-sign template we obtain the following three-cell CNN ($M = 3$, $N = 1$, $C = 1$, $R = 1$, and $u_{kl} = 0$):

$$\begin{aligned} {}_0D_t^{q_1}x_1(t) &= -x_1(t) + p_1 f(x_1(t)) - sf(x_2(t)) - sf(x_3(t)), \\ {}_0D_t^{q_2}x_2(t) &= -x_2(t) - sf(x_1(t)) + p_2 f(x_2(t)) - rf(x_3(t)), \\ {}_0D_t^{q_3}x_3(t) &= -x_3(t) - sf(x_1(t)) + rf(x_2(t)) + p_3 f(x_3(t)), \end{aligned} \quad (5.93)$$

where $p_1 > 1$, $p_2 > 1$, $p_3 \geq 1$, $r > 0$, and $s > 0$ are the CNN parameters, q_1, q_2, and q_3 are the derivative orders for each cell (related to the capacitor order m).

Let us assume that we have the three-identical-cell CNN described by Eqs. (5.93), with fractional commensurate order $q_1 = q_2 = q_3 = 0.99$ (orders of real analog capacitors). We can use the relations (2.53) and (2.54) to derive a numerical solution of the fractional-order CNN described by (5.93) as follows:

$$\begin{aligned} x_1(t_k) &= (-x_1(t_{k-1}) + p_1 f(x_1(t_{k-1})) - sf(x_2(t_{k-1})) - sf(x_3(t_{k-1})))\,h^{q_1} \\ &\quad - \sum_{j=v}^{k} c_j^{(q_1)} x_1(t_{k-j}), \\ x_2(t_k) &= (-x_2(t_{k-1}) - sf(x_1(t_k)) + p_2 f(x_2(t_{k-1})) - rf(x_3(t_{k-1})))\,h^{q_2} \\ &\quad - \sum_{j=v}^{k} c_j^{(q_2)} x_2(t_{k-j}), \\ x_3(t_k) &= (-x_3(t_{k-1}) - sf(x_1(t_k)) + rf(x_2(t_k)) + p_3 f(x_3(t_{k-1})))\,h^{q_3} \\ &\quad - \sum_{j=v}^{k} c_j^{(q_3)} x_3(t_{k-j}), \end{aligned} \quad (5.94)$$

where a nonlinear function $f(.)$ is defined by (5.91), T_{sim} is the simulation time, $k = 1, 2, 3, \ldots, N$, for $N = [T_{sim}/h]$, and $(x_1(0), x_2(0), x_3(0))$ is the start point (initial conditions).

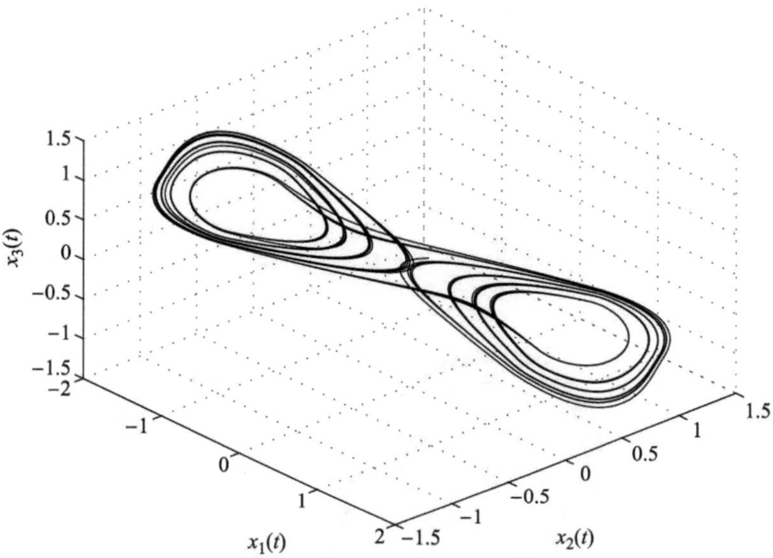

Fig. 5.58 State space trajectory of the CNN (5.93) for the parameters: $p_1 = 1.24$, $p_2 = 1.1$, $p_3 = 1$, $s = 3.21$, $r = 4.4$, orders $q_1 = q_2 = q_3 = 0.99$, with initial conditions: $(x_1(0), x_2(0), x_3(0)) = (0.1, 0.1, 0.1)$.

The simulation result for time step $h = 0.005$ and simulation time $T_{sim} = 100\,s$ is depicted in Fig. 5.58. The result shows the chaotic double scroll attractor of the three-cell fractional-order CNNs (5.93) for the parameters $p_1 = 1.24$, $p_2 = 1.1$, $p_3 = 1$, $s = 3.21$, $r = 4.4$, and system orders $q_1 = 0.99$, $q_2 = 0.99$, and $q_3 = 0.99$ with the initial conditions: $(x_1(0), x_2(0), x_3(0)) = (0.1, 0.1, 0.1)$.

5.16 Fractional-Order Volta's System

The system was discovered by Volta – a student at the Department of Physics, Genova University, in 1984, when writing his thesis with Prof. Antonio Borsellino and Dr. Francisco Fu Arcardi.

Volta's system is described by the system of state differential equations (Hao, 1989):

$$
\begin{aligned}
\dot{x}(t) &= -x(t) - 5y(t) - z(t)y(t), \\
\dot{y}(t) &= -y(t) - 85x(t) - x(t)z(t), \\
\dot{z}(t) &= 0.5z(t) + x(t)y(t) + 1.
\end{aligned}
\tag{5.95}
$$

The Lyapunov exponents (LE) of the system (5.95) computed according to the method described in the work (Wolf et al., 1985) are: $LE_1 = 0.064979$, $LE_2 = -1.0708$, and $LE_3 = -1.4936$ for initial values (8, 2, 1).

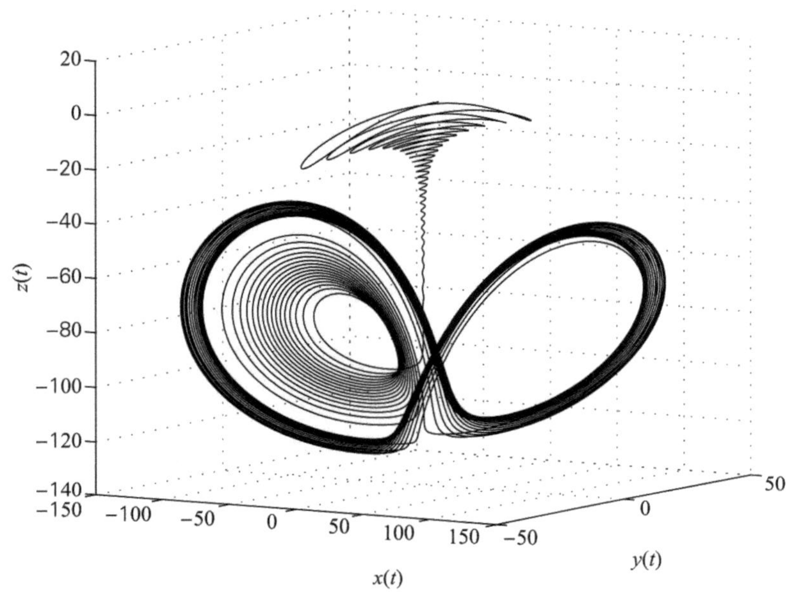

Fig. 5.59 Chaotic attractor of Volta's system (5.95) projected into 3D state space for initial conditions $(x(0),y(0),z(0)) = (8,2,1)$ and $T_{sim} = 20\,s$.

Fig. 5.60 shows the time histories of variables $x(t)$, $y(t)$, and $z(t)$ of the system (5.95), for $T_{sim} = 10\ s$. In Fig. 5.59 is depicted a phase trajectory in 3D state-space of Volta's system (5.95) for $T_{sim} = 20\,s$ starting from $(x(0),y(0),z(0)) = (8,2,1)$.

In Fig. 5.61 are shown the phase trajectories, starting from $(x(0),y(0),z(0)) = (8,2,1)$, and projected into 2D phase planes, respectively.

Obviously, if the Laypunov exponent LE_1 is positive and if we observe strange attractors in Fig. 5.59 and Fig. 5.61, the system (5.95) has chaotic behavior toward initial values $(x(0),y(0),z(0)) = (8,2,1)$.

We can generalize Volta's system (5.95) to the following form:

$$\begin{aligned} \frac{dx(t)}{dt} &= -x(t) - ay(t) - z(t)y(t), \\ \frac{dy(t)}{dt} &= -y(t) - bx(t) - x(t)z(t), \\ \frac{dz(t)}{dt} &= cz(t) + x(t)y(t) + 1. \end{aligned} \tag{5.96}$$

The Jacobian matrix of the system (5.96), evaluated at the equilibrium $E^* = (x^*,y^*,z^*)$, is given by

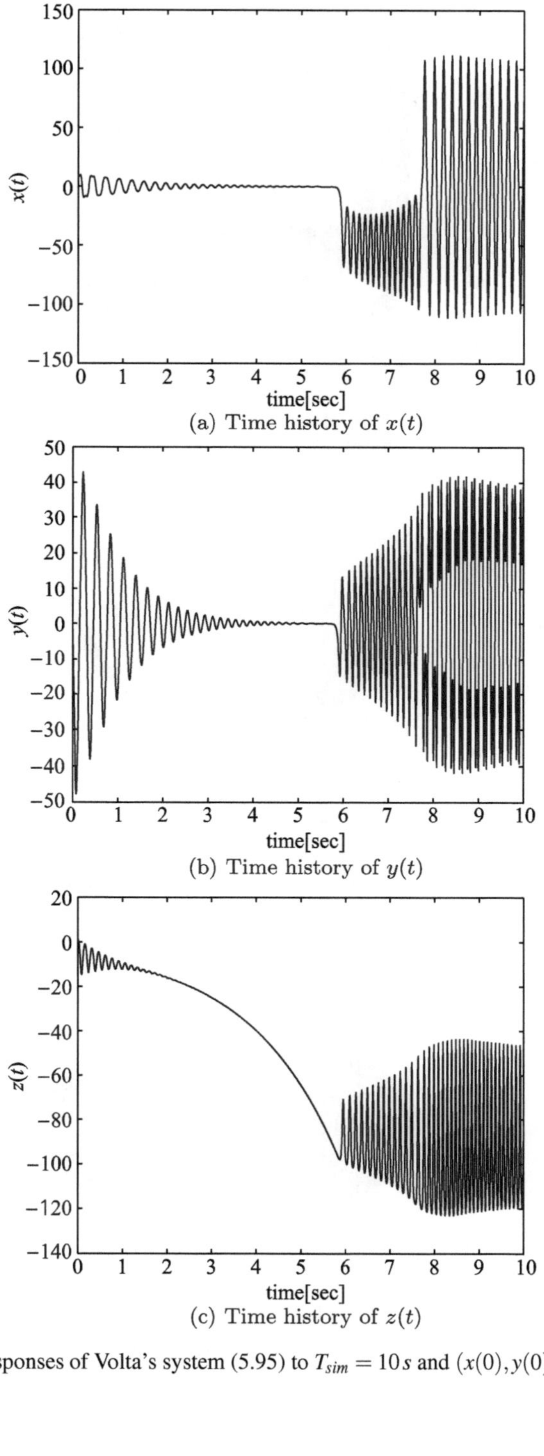

(a) Time history of $x(t)$

(b) Time history of $y(t)$

(c) Time history of $z(t)$

Fig. 5.60 Time responses of Volta's system (5.95) to $T_{sim} = 10\,s$ and $(x(0), y(0), z(0)) = (8, 2, 1)$.

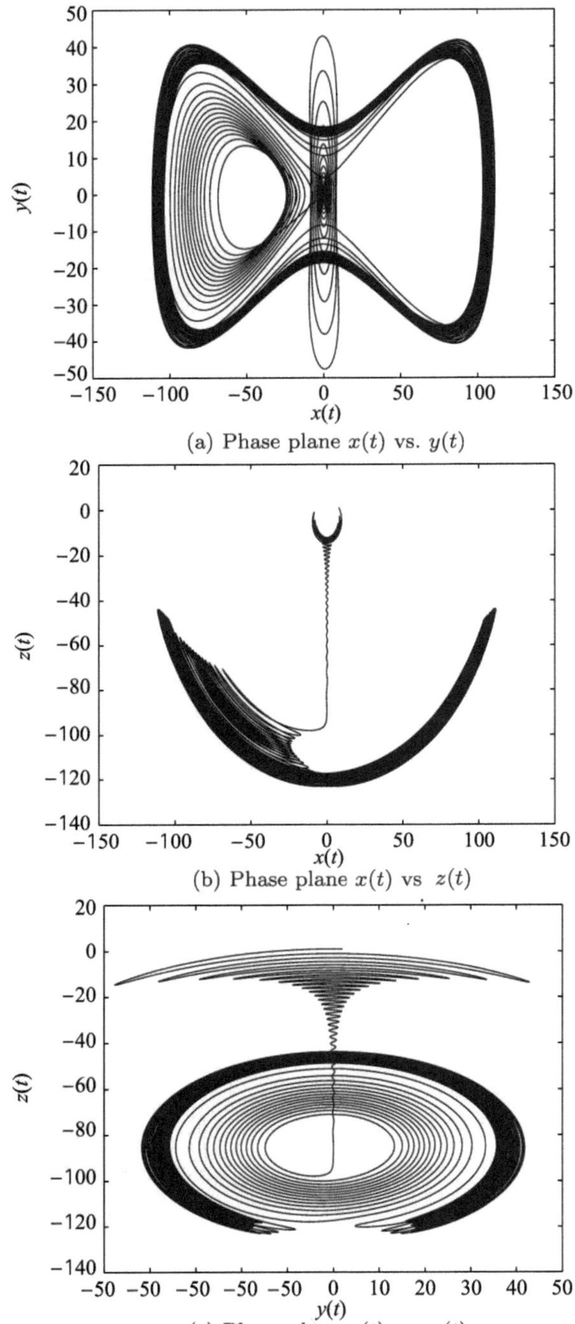

(a) Phase plane $x(t)$ vs. $y(t)$

(b) Phase plane $x(t)$ vs $z(t)$

(c) Phase plane $y(t)$ vs. $z(t)$

Fig. 5.61 Phase plane projections of Volta's system (5.95) for $T_{sim} = 20\,s$.

$$\mathbf{J} = \begin{bmatrix} -1 & -a-z^* & -y^* \\ -b-z^* & -1 & -x^* \\ y^* & x^* & c \end{bmatrix}. \tag{5.97}$$

When $(a,b,c) = (5,85,0.5)$, Volta's system shows chaotic behavior (Fig. 5.59 and Fig. 5.61). For these system parameters, Volta's system has three quilibria $E_1 = (0;0;-2)$, $E_2 = (-57.6282;-0.7202;-85.0124)$, $E_3 = (57.6282;0.7202;-85.0124)$ and their corresponding eigenvalues are: $\lambda_1 \approx 14.7797$, $\lambda_2 \approx -16.7797$, $\lambda_3 = 0.5$ for E_1, $\lambda_1 = -2$, $\lambda_{2,3} \approx 0.25 \pm 57.6322j$ for E_2, and $\lambda_1 \approx -10.6861$, $\lambda_2 \approx 11.18617$, $\lambda_3 - 2$ for E_3.

Hence, the equilibria E_1 and E_3 are unstable saddle points. The equilibrium E_2 is a saddle-focus point. According to the stability conditions (4.40), where $q = 1$, we have eigenvalues for equilibria E_1, E_2 and E_3 in the unstable region and therefore we can confirm the chaotic behavior of Volta's systems (5.95) for the initial conditions $(x(0),y(0),z(0)) = (8,2,1)$.

Now, we consider the fractional-order Volta's system, where integer-order derivatives are replaced by fractional-order ones. The mathematical description of the fractional-order chaotic system is expressed as (Petráš, 2009a):

$$\begin{aligned} {}_0D_t^{q_1}x(t) &= -x(t) - ay(t) - z(t)y(t), \\ {}_0D_t^{q_2}y(t) &= -y(t) - bx(t) - x(t)z(t), \\ {}_0D_t^{q_3}z(t) &= cz(t) + x(t)y(t) + 1, \end{aligned} \tag{5.98}$$

where q_1, q_2, and q_3 are the derivative orders. The total order of the system is $\bar{q}$=(q_1, q_2, q_3).

For numerical solution of the chaotic system (5.98) we use the relationship (2.53), which leads to approximation in FIR form. By setting $N = [T_{sim}/h]$, we have

$$\begin{aligned} x(t_k) &= \left(-x(t_{k-1}) - ay(t_{k-1}) - z(t_{k-1})y(t_{k-1})\right)h^{q_1} - \sum_{j=v}^{k} c_j^{(q_1)} x(t_{k-j}), \\ y(t_k) &= \left(-y(t_{k-1}) - bx(t_k) - x(t_k)z(t_{k-1})\right)h^{q_2} - \sum_{j=v}^{k} c_j^{(q_2)} y(t_{k-j}), \\ z(t_k) &= \left(cz(t_{k-1}) + x(t_k)y(t_k) + 1\right)h^{q_3} - \sum_{j=v}^{k} c_j^{(q_3)} z(t_{k-j}), \end{aligned} \tag{5.99}$$

where T_{sim} is the simulation time, $k = 1,2,3,\ldots,N$ and $(x(0),y(0),z(0))$ is the start point (initial conditions). The binomial coefficients $c_j^{(q_i)}$, $\forall i$ are calculated according to the relation (2.54).

When we assume the same orders of derivatives in state equations (5.98), i.e. $q_1 = q_2 = q_3 \equiv q$, we get a commensurate-order system. According to condition (4.42) it is determined that the commensurate order q of derivatives has to be

$q > 0.99$. It means that for system parameters $(a,b,c) = (5,85,0.5)$, only for integer order $q = 1$, chaos can be observed. If we would like to go to the fractional (commensurate) order, we have to change the system parameters, e.g. for system parameters $(a,b,c) = (19,11,0.73)$, chaos can be observed if $q > 0.977$. For these parameter sets, Volta's system has three equilibria $E_1 = (0;0;-1.3698)$, $E_2 = (-1.26310;-10.26032;-19.12310)$, $E_3 = (1.26310;10.26032;-19.12310)$ and their corresponding eigenvalues are $\lambda_1 \approx 12.0299$, $\lambda_2 \approx -14.0299$, $\lambda_3 - 0.73$ for E_1, $\lambda_1 = -2$, $\lambda_{2,3} \approx 0.3650 \pm 10.3313j$ for E_2, and $\lambda_1 \approx -7.2088$, $\lambda_2 \approx 7.93883$, $\lambda_3 = -2$ for E_3.

Hence, the equilibria E_1 and E_3 are saddle points and the equilibrium E_2 is saddle-focus point. The characteristic equation evaluated at equilibrium E_1 is

$$\lambda^{294} + 127/100\lambda^{196} - 170.2406\lambda^{98} + 123.2098 = 0,$$

with unstable roots $\lambda_1 \approx 0.9968$ and $\lambda_2 \approx 1.0257$.

The characteristic equation evaluated at equilibrium E_2 is

$$\lambda^{294} + 127/100\lambda^{196} + 105.40980\lambda^{98} + 213.7396 = 0,$$

with unstable roots $\lambda_{1,2} \approx 1.0240 \pm 0.0160j$, becasue $|\arg(\lambda_{1,2})| = 0.0157 < \pi/2m$, where $m = 100$ (LCM of orders denominator).

The characteristic equation evaluated at equilibrium E_3 is

$$\lambda^{294} + 127/100\lambda^{196} - 58.6898\lambda^{98} - 114.45960 = 0,$$

with unstable root $\lambda_1 \approx 1.0214$.

According to the stability conditions (4.40), where $q = 0.98$, we have eigenvalues of the equilibrium points E_1, E_2 and E_3 in the unstable region and therefore we can confirm the chaotic behavior of Volta's systems (5.98) for the initial conditions $(8,2,1)$. Instability measure is 0.0157. It means that commensurate fractional-order Volta's system is chaotic.

In Fig. 5.62 is shown the chaotic behavior toward fractional-order system (5.98), where system parameters are $(a,b,c) = (19,11,0.73)$, commensurate order of the derivatives is $q = 0.98$, the initial conditions are $(x(0),y(0),z(0)) = (8,2,1)$ for simulation time $T_{sim} = 20\,s$ and time step $h = 0.0005$.

When we assume the different orders of derivatives in state equations (5.98), i.e. $q_1 \neq q_2 \neq q_3$, we get a general incommensurate-order system. There is no exact condition for determining the orders to obtain chaotic behavior of the system. We experimentally found the following orders (Petráš, 2010): $q_1 = 0.89$, $q_2 = 1.10$, and $q_3 = 0.91$ for system parameters $(a,b,c) = (5,85,0.5)$.

The stability can be investigated according to characteristic equation (4.41). For the above derivative orders and the system parameters, and for the Jacobian matrix (5.97) evaluated at the equilibrium points E^*, Equation (4.41) becomes

$$\det(\mathrm{diag}([\lambda^{89}\ \lambda^{110}\ \lambda^{91}]) - \mathbf{J}) = 0, \tag{5.100}$$

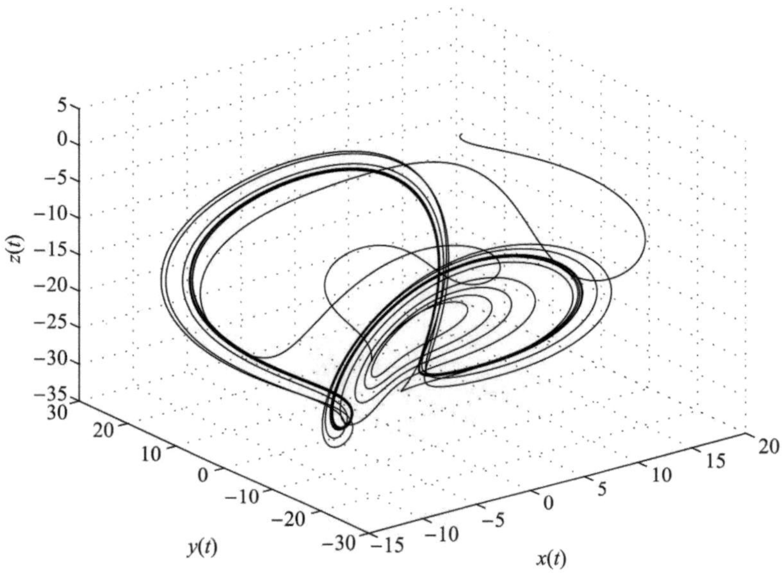

Fig. 5.62 Chaotic attractor of Volta's system (5.98) projected into 3D state space for initial conditions $(x(0),y(0),z(0)) = (8,2,1)$, parameters $(a,b,c) = (19,11,0.73)$, orders $(q_1,q_2,q_3) \equiv (q = 0.98)$ and $T_{sim} = 20\,s$.

where the LCM is $m = 100$. The characteristic equation (5.100) evaluated at equilibrium E_1 is

$$\lambda^{290} + \lambda^{201} - 1/2\lambda^{199} + \lambda^{180} - 1/2\lambda^{110} - 248\lambda^{91} - 1/2\lambda^{89} + 124 = 0,$$

with unstable roots $\lambda_1 \approx 1.0274$ and $\lambda_2 \approx 0.9924$.

The characteristic equation evaluated at equilibrium E_2 is

$$\lambda^{290} - 1/2\lambda^{199} + \lambda^{180} + 3320.5186\lambda^{89} + \lambda^{201} + \\ 0.01874\lambda^{110} - 0.127x10^{-28}\lambda^{91} + 6643.0748 = 0,$$

with unstable roots $\lambda_{1,2} \approx 1.0411 \pm 0.0161j$, becasue $|\arg(\lambda_{1,2})| = 0.0155 < \pi/2m$, where $m = 100$ (LCM of orders denominator).

The characteristic equation evaluated at equilibrium E_3 is

$$\lambda^{290} - 1/2\lambda^{199} + \lambda^{180} - 0.5186\lambda^{89} + \lambda^{201} - \\ 120.0187\lambda^{110} + 0.333 \times 10^{-29}\lambda^{91} - 239.0748 = 0,$$

which has unstable root $\lambda_1 = 1.0270$.

Because the system has unstable roots, the system satisfies the necessary condition for exhibiting chaotic attractor. Instability measure is 2.137×10^{-4}.

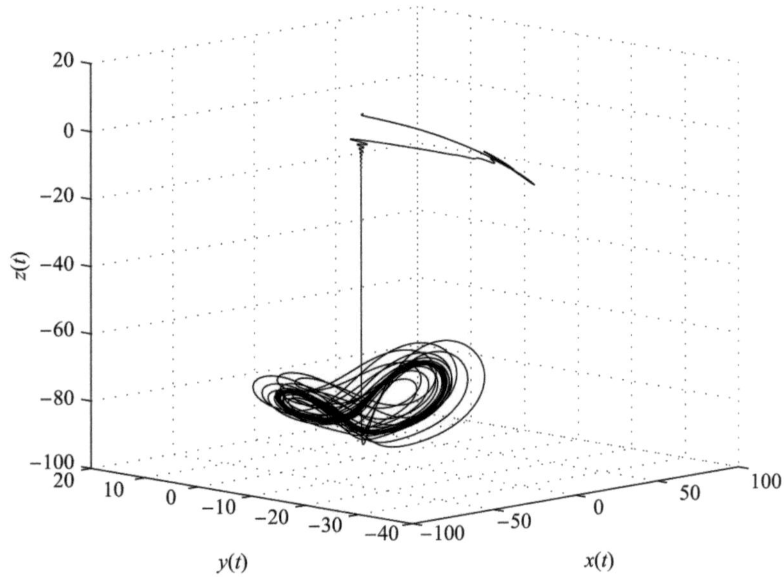

Fig. 5.63 Chaotic attractor of Volta's system (5.98) projected into 3D state space for initial conditions $(x(0),y(0),z(0)) = (8,2,1)$, parameters $(a,b,c) = (5,85,0.5)$, orders $(q_1,q_2,q_3) \equiv \bar{q} \in (0.89,1.10,0.91)$ and $T_{sim} = 20s$.

In Fig. 5.63 is shown the chaotic behavior for fractional-order chaotic system (5.98), where system parameters are $(a,b,c) = (5,85,0.5)$, incommensurate orders of the derivatives are: $q_1 = 0.89$, $q_2 = 1.10$, and $q_3 = 0.91$, and the initial conditions are $(x(0),y(0),z(0)) = (8,2,1)$ for the simulation time $T_{sim} = 20s$ and time step $h = 0.0005$. As we can see, behavior of the fractional-order Volta's system is still chaotic because we have observed double-scroll attractor (Tavazoei and Haeri, 2007b) and total order of the system is $\bar{q} = 2.9$.

The state equations of the fractional-order Volta's chaotic system (5.98) with parameters $(a,b,c) = (5,85,0.5)$ are given by using the integration operation and the properties (2.50), and (2.51) and have form:

$$
\begin{aligned}
x(t) &= {}_0D_t^{1-q_1}\left(\int_0^t [-x(t)-5y(t)-z(t)y(t)]dt\right),\\
y(t) &= {}_0D_t^{1-q_2}\left(\int_0^t [-y(t)-85x(t)-x(t)z(t)]dt\right),\\
z(t) &= {}_0D_t^{1-q_3}\left(\int_0^t [0.5z(t)+x(t)y(t)+1]dt\right).
\end{aligned}
\tag{5.101}
$$

The system model developed from the state equations (5.101) for system parameters $(a,b,c) = (5,85,0.5)$ by using the Matlab/Simulink environment is depicted in Fig. 5.64.

The simulation results for simulation time $T_{sim} = 20$ s obtained from model (5.101) for real order $q_1 = 0.93$, $q_2 = 0.99$, and $q_3 = 0.98$ are dipicted in Fig. 5.65. As we can see, behavior of the fractional-order Volta's system is chaotic because we have observed double scroll attractor in $x-y$ plane (Tavazoei and Haeri, 2007b) and the total order of the system is $\bar{q} = 2.9$.

In Fig. 5.65 are shown the phase trajectories of the Simulink system model, starting from (x(0), y(0), z(0))=(8, 2, 1), and projected into 2D phase planes, respectively.

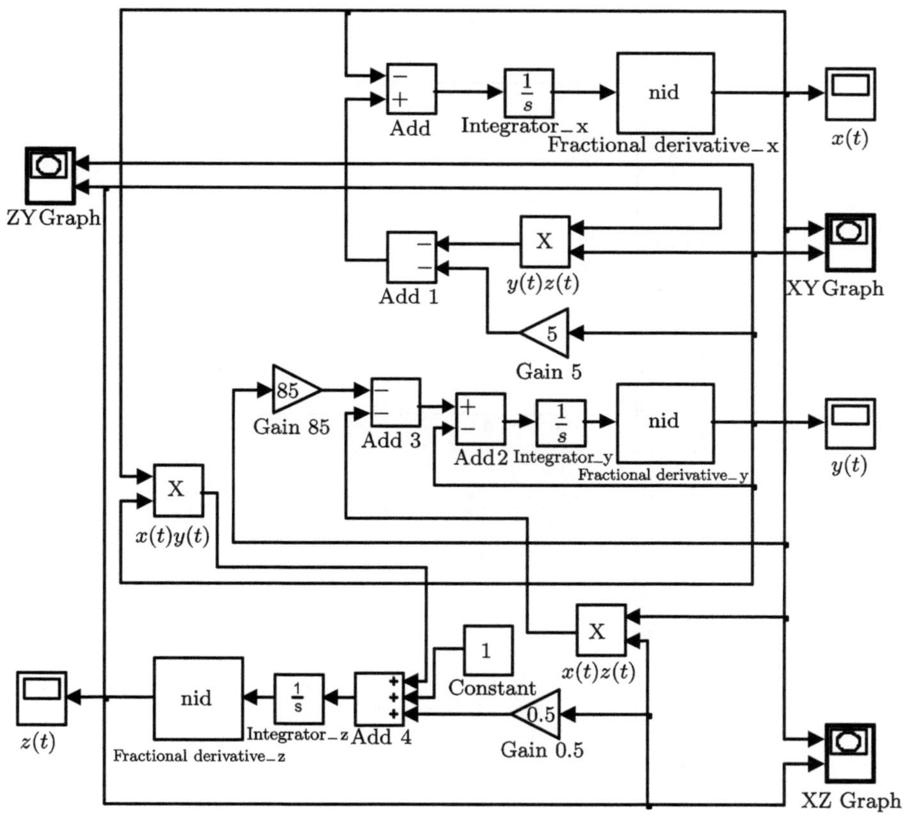

Fig. 5.64 Matlab/Simulink block diagram (model) for fractional order Volta's system (5.101).

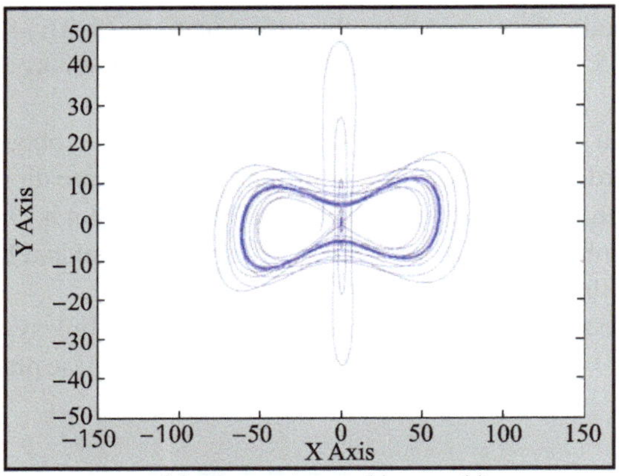

(a) Phase plane $x(t)$ vs. $y(t)$

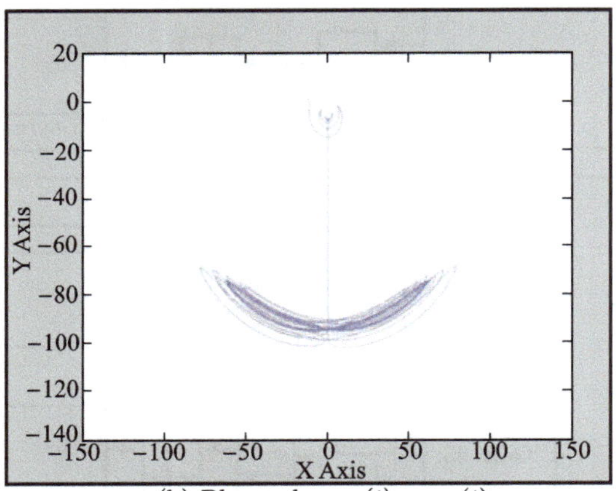

(b) Phase plane $x(t)$ vs. $z(t)$

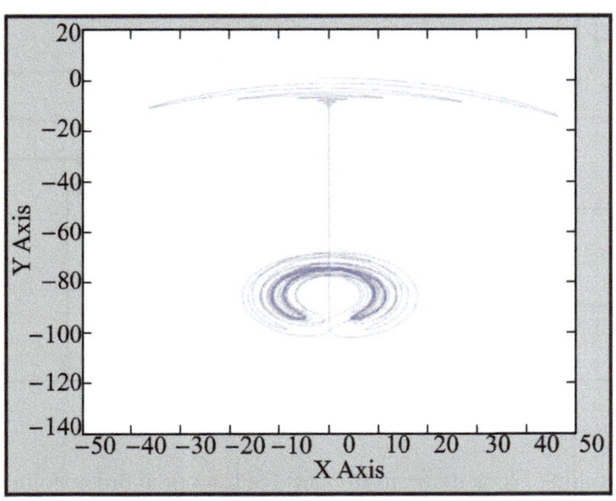

(c) Phase plane $y(t)$ vs. $z(t)$

Fig. 5.65 Phase plane projections of fractional-order Volta's system (5.101) for $T_{sim} = 20\,s$.

References

Ahmed E., El-Sayed A. M. A. and El-Saka H. A. A., 2007, Equilibrium points, stability and numerical solutions of fractional-order predator-prey and rabies models, *J. Math. Anal. Appl.*, **325**, 542–553.

Arena P., Caponetto R., Fortuna L. and Porto D., 1998, Bifurcation and chaos in noninteger order cellular neural networks, *International Journal of Bifurcation and Chaos*, **8**, 1527–1539.

Arena P., Caponetto R., Fortuna L. and Porto D., 2000, *Nonlinear Noninteger Order Circuits and Systems – An Introduction*, World Scientific, Singapore.

Barbosa R. S., Machado J. A. T, Vinagre B. M. and Calderón A. J., 2007, Analysis of the Van der Pol oscillator containing derivatives of fractional order, *Journal of Vibration and Control*, **13**, 1291–1301.

Bartissol P. and Chua L. O., 1988, The double hook, *IEEE Transactions on Circuits and Systems*, **35**, 1512–1522.

Biey M., Checco P. and Gilli M., 2003, Bifurcation and chaos in cellular neural networks, *Journal of Circuits, Systems, and Computers*, **12**, 417–433.

Caponetto R., Dongola G., Fortuna L. and Petráš I., 2010, *Fractional Order Systems: Modeling and Control Applications*, World Scientific, Singapore.

Carlson G. E. and Halijak C. A., 1963, Approximation of fractional capacitors $(1/s)^{1/n}$ by a regular Newton process, *Proc. of the Sixth Midwest Symposium on Circuit Theory*, May 6–7, Madison, Wisconsin.

Cruz J. M., 1993, An IC chip of Chua's circuit, *IEEE Transactions on Circuits and Systems –II: Analog and Digital Signal Processing*, **40**, 614–625.

Chen J. H. and Chen W. Ch., 2008, Chaotic dynamics of the fractionally damped van der Pol equation, *Chaos, Solitons and Fractals*, **35**, 188–198.

Chen W. Ch., 2008, Nonlinear dynamic and chaos in a fractional-order financial system, *Chaos, Solitons and Fractals*, **36**, 1305–1314.

Chua L. O., 1971, Memristor – the missing circuit element, *IEEE Transaction on Circuit Theory*, **CT-18**, 507–519.

Chua L. O., Komuro M. and Matsumoto T., 1986, The double scroll family, *IEEE Trans. on Circuit and Systems*, **CAS-33**, 1073–1118.

Chua L. O. and Roska T., 1993, The CNN paradigm, *IEEE Transactions on Circuits and Systems–I: Fundamental Theory and Applications*, **40**, 147–157.

Chua L. O., Wu Ch. W., Huang A. and Zhong G. Q., 1993, A universal circuit for studying and generating chaos – part I: Routes to chaos, *IEEE Transactions on Circuits and Systems –I: Fundamental Theory and Applications*, **40**, 732–744.

Deng W. H. and Li C. P., 2005, Chaos synchronization of the fractional Lü system, *Physica A*, **353**, 61–72.

Deregel P., 1993, Chua's oscillator: A zoo of attractors, *Journal of Circuits, Systems, and Computers*, **3**, 309–359.

Gao X. and Yu J., 2005, Chaos in the fractional order periodically forced complex Duffing's oscillators, *Chaos, Solitons and Fractals*, **24**, 1097–1104.

Ge Z. M. and Hsu M. Y., 2007, Chaos in a generalized van der Pol system and in its fractional order system, *Chaos, Solitons and Fractals*, **33**, 1711–1745.

Ge Z. M. and Ou Ch. Y., 2007, Chaos in a fractional order modified Duffing system, *Chaos, Solitons and Fractals*, **34**, 262–291.

Gejji V. D. and Bhalekar S., 2010, Chaos in fractional ordered Liu system, *Computers & Mathematics with Applications*, **59**, 1117–1127.

Genesio R. and Tesi A., 1992, Harmonic balance methods for the analysis of chaotic dynamics in nonlinear systems, *Automatica*, **28**, 531–548.

Guo L. J., 2005, Chaotic dynamics and synchronization of fractional-order Genesio-Tesi systems, *Chinese Physics*, **14**, 1517–1521.

Hao B., 1989, *Elementary Symbolic Dynamics and Chaos in Dissipative Systems*, World Scientific, Singapore.

Hartley T. T., Lorenzo C. F. and Qammer H. K., 1995, Chaos on a fractional Chua's system, *IEEE Trans. Circ. Syst. Fund. Theor. Appl.*, **42**, 485–490.

Itoh M. and Chua L. O., 2008, Memristor oscillation, *International Journal of Bifurcation and Chaos*, **18**, 3183–3206.

Itoh M. and Chua L. O., 2009, Memristor cellular automata and memristor discrete-time cellular neural networks, *International Journal of Bifurcation and Chaos*, **19**, 3605–3656.

Kennedy M. P., 1992, Robust OP AMP realization of Chua's circuit, *Frequenz*, **46**, 66–80.

Kennedy M. P., 1993a, Three steps to chaos – part I: Evolution, *IEEE Transactions on Circuits and Systems –I: Fundamental Theory and Applications*, **40**, 640–656.

Kennedy M. P., 1993b, Three steps to chaos – part II: A Chua's circuit primer, *IEEE Transactions on Circuits and Systems –I: Fundamental Theory and Applications*, **40**, 657–674.

Leipnik R. B. and Newton T. A., 1981, Double strange attractors in rigid body motion with linear feedback control, *Physics Letters A*, **86**, 63–67.

Li C. and Chen G., 2004, Chaos and hyperchaos in the fractional-order Rössler equations, *Physica A*, **341**, 55–61.

Li C. and Yan J., 2007, The synchronization of three fractional differential systems, *Chaos, Solitons and Fractals*, **32**, 751–757.

Liu C., Liu L. and Liu T., 2009, A novel three-dimensional autonomous chaos system, *Chaos, Solitons and Fractals*, **39**, 1950–1958.

Lu J. G., 2005, Chaotic dynamics and synchronization of fractional-order Arneodo's systems, *Chaos, Solitons and Fractals*, **26**, 1125–1133.

Lu J. and Chen G., 2002, A new chaotic attractor coined, *International Journal of Bifurcation and Chaos*, **12**, 659–661.

Lu J. G. and Chen G., 2006, A note on the fractional-order Chen system, *Chaos, Solitons and Fractals*, **27**, 685–688.

Ma J. H. and Chen Y. S., 2001a, Study for the bifurcation topological structure and the global complicated character of a kind of nonlinear finance system (I), *Applied Mathematics and Mechanics*, **22**, 1240–1251.

Ma J. H. and Chen Y. S., 2001b, Study for the bifurcation topological structure and the global complicated character of a kind of nonlinear finance system (II), *Applied Mathematics and Mechanics*, **22**, 1375–1382.

Matsumoto T., 1984, A chaotic attractor from Chua's circuit, *IEEE Transactions on Circuit and Systems*, **CAS-31**, 1055–1058.

Oustaloup A., Levron F., Mathieu B. and Nanot F. M., 2000, Frequency-band complex noninteger differentiator: characterization and synthesis, *IEEE Trans. on Circuits and Systems I: Fundamental Theory and Applications I*, **47**, 25–39.

Petráš I., 2006, A Note on the fractional-order cellular neural networks, *Proc. of the International Joint Conference on Neural Networks*, July 16-21, Vancouver, Canada, 1021–1024.

Petráš I., 2008, A note on the fractional-order Chua's system, *Chaos, Solitons and Fractals*, **38**, 140–147.

Petráš I., 2009a, Chaos in the fractional-order Volta's system: modeling and simulation, *Nonlinear Dynamics*, **57**, 157–170.

Petráš I., 2009b, *Fractional-Order Chaotic Systems*, Reprocentrum, Faculty of BERG, Technical University of Kosice.

Petráš I. and Bednárová D., 2009, Fractional – order chaotic systems, *Proc. of the IEEE Conference on Emerging Technologies & Factory Automation, ETFA 2009*, September 22–25, Palma de Mallorca, Spain.

Petráš I., 2010, A note on the fractional-order Volta's system, *Commun. Nonlinear. Sci. Numer. Simulat.*, **15**, 384–393.

Pivka L., Wu C. W. and Huang A., 1994, Chua's Oscillator: A Compendium of Chaotic Phenomena, *Journal of the Franklin Institute*, **33/B**, 705–741.

Podlubny I., 1999, *Fractional Differential Equations*, Academic Press, San Diego.

Samardzija N. and Greller L. D., 1988, Explosive route to chaos through a fractal torus in a generalized Lotka-Volterra model, *Bulletin of Mathematical Biology*, **50**, 465–491.

Schafer I. and Kruger K., 2008, Modelling of lossy coils using fractional derivatives, *J. Phys. D: Appl. Phys.*, **41**, 1–8.

Sheu L. J., Chen H. K., Chen J. H., Tam L. M., Chen W. C., Lin K. T. and Kang Y., 2008, Chaos in the Newton–Leipnik system with fractional order, *Chaos, Solitons and Fractals*, **36**, 98–103.

Tavazoei M. S. and Haeri M., 2007a, Unreliability of frequency-domain approximation in recognising chaos in fractional-order systems, *IET Signal Proc.*, **1**, 171–181.

Tavazoei M. S. and Haeri M., 2007b, A necessary condition for double scroll attractor existence in fractional – order systems, *Physics Letters A*, **367**, 102–113.

Valerio D., 2005, Toolbox `ninteger` for Matlab, v. 2.3 (September 2005), http://web.ist.utl.pt/duarte.valerio/ninteger/ninteger.htm.

Westerlund S. and Ekstam L., 1994, Capacitor theory, *IEEE Trans. on Dielectrics and Electrical Insulation*, **1**, 826–839.

Westerlund S., 2002, *Dead Matter Has Memory*! Causal Consulting, Kalmar, Sweden.

Wolf A., Swift J. B., Swinney H. L. and Vastano J. A., 1985, Determining Lyapunov exponents from a time series, *Physica D*, **16**, 285–317.

Zaslavsky G. M., 2005, *Hamiltonian Chaos and Fractional Dynamics*, Oxford University Press, Oxford.

Zhong G. Q., 1994, Implementation of Chua's circuit with a cubic nonlinearity, *IEEE Transactions on Circuits and Systems –I: Fundamental Theory and Applications*, **41**, 934–941.

Zou F. and Nossek J. A., 1993, Bifurcation and chaos in cellular neural networks, *IEEE Transactions on Circuits and Systems-I: Fundamental Theory and Applications*, **40**, 166–173.

Zhou T., Tang Y. and Chen G., 2004, Chen's attractor exists, *International Journal of Bifurcation and Chaos*, **14**, 3167–3177.

Zhu H., Zhou S. and Zhang J., 2009, Chaos and synchronization of the fractional-order Chua's system, *Chaos, Solitons and Fractals*, **39**, 1595–1603.

Chapter 6
Control of Fractional-Order Chaotic Systems

6.1 Preliminary Considerations

Control of nonlinear systems, especially chaotic systems, was the subject of intensive studies in the last few decades. As noted (Andrievskii and Fradkov, 2003, 2004), several thousand publications have appeared over the recent decade due to the fact that chaotic behavior was discovered in numerous systems in mechanics, laser and radio physics, hydrodynamics, chemistry, biology and medicine, electronic circuits, economical systems, etc. For this reason a natural question arises: How can we control chaotic systems?

The first important thing is that we need the mathematical formulation of chaotic processes by the basic models of the chaotic systems that are used. The most popular mathematical models used in the literature on control of chaos are represented by the systems of ordinary differential equations. In some works we can also find discrete models defined by difference state equations. The second important thing is the formulation of the problems of control ing chaotic processes. An important type of problems of controling chaotic processes is represented by the modification of the attractors, for example, transformation of chaotic oscillations into periodic state and so on.

It could be a problem of stabilization of unstable periodic solutions. The problem lies in determining the control function as either the open-loop control action, or the state feedback, or the output feedback satisfying the control objective. Stabilization of an unstable equilibrium is a special case.

The second class includes the control problems of excitation or generation of chaotic oscillations. These problems are also called chaotization or anticontrol. They arise in the circumst ances where chaotic motion is the desired behavior of the system. These problems are characterized by the fact that the trajectory of the system phase vector is not predetermined, is unknown, or is of no consequence for attaining the objective.

The third important class of the control objectives corresponds to the problems of synchronization or, more precisely, controllable synchronization as opposed to the autosynchronization. Numerous publications on control of synchronization of chaotic processes and its application in the data transmission systems appeared in the 1990's. In the general case, by the synchronization is meant the coordinated variation of the states of two or more systems or, possibly, coordinated variation of some of their characteristics such as oscillation frequencies.

Let us take a look at the synchronization more closely. Several methods can be used for synchronization of chaotic systems. In this paragraph we will mention three well-known methods. If chaos synchronization is achieved by drive-response systems, the instability measure is negative. That means the system is not chaotic.

The first method is the **M**aster-**S**lave (or drive-response) configuration scheme of two autonomous n-dimensional fractional-order chaotic systems (Peng, 2007):

$$\begin{cases} \mathrm{M}: \dfrac{d^{\alpha}x}{dt^{\alpha}} = f(x), \\ \mathrm{S}: \dfrac{d^{\alpha}\tilde{x}}{dt^{\alpha}} = f(\tilde{x}) + C(x-\tilde{x}), \end{cases} \tag{6.1}$$

where $\alpha = (\alpha_1, \alpha_2, \ldots, \alpha_n) \in \mathrm{R}^n$, $\alpha_i > 0$, is the fractional order and the systems are chaotic. C is a coupling matrix. For simplicity, let C have the form: $C = \mathrm{diag}(d_1, d_2, \ldots, d_n)$, where $d_i \geq 0$. The error is $e = x - \tilde{x}$ and the aim of the synchroniziation is to design the coupling matrix so that $||e(t)|| \to 0$ as $t \to +\infty$.

The second method is the method for constructing the drive-response configuration which was introduced by Pecora and Carroll (Pecora and Carroll, 1990), known as a PC. Let us build a PC drive-response configuration in which a drive system is given by the fractional-order system (with three state variables: x, y, z) and a response system is given by the subspace containing the (x, y) variables. Then we can use the chaotic signal z to drive the response subsystem (Deng and Li, 2005).

The third method is the synchronization via active-passive decomposition method (APD). Let us build an APD drive-response configuration with a drive system given by the chaotic system and with a response system given by its replica. Then we can take $s(t)$ as a drive signal (Li et al., 2006).

Chaos synchronization and its potential application to secure communications have attracted much attention from scholars engaged in various disciplines in science and engineering since the pioneering work of Pecora and Carroll (Pecora and Carroll, 1990). In this section, we briefly discuss the chaos synchronization methods between the chaotic fractional-order systems an we can also mention method via master-slave configuration with linear coupling (Zhu et al., 2009).

The fractional calculus techniques as, for example, a fractional-differentiator-based controller of a fractional-integrator-based controller can also be used (Tavazoei et al., 2009). Both of them are particular cases of the fractional-order controllers described in Section. 3.3, which are more flexible than classical ones and give better results of the control performances (Caponetto et al., 2010; Chen et al., 2009). A different type of controller was proposed (Hosseinnia et al., 2010).

A detailed review of several applications, where chaos was controlled in systems of various disciplines such as mechanics, physics (control of turbulence, lasers, and plasma), chemistry and raw material processing, biology and ecology (control of population), economics, medicine, space engineering (control of chaotic spacecraft angular oscillations), electrical and electronic systems, communication systems, and information systems, was decribed (Andrievskii and Fradkov, 2004).

6.2 A Survey of Control Strategies

In the paper (Andrievskii and Fradkov, 2003) were collected and presented several methods used for the control of chaotic processes. The authors considered the classical integer-order chaotic systems but in general we can use those methods for the fractional-order chaotic systems as well. In addition, some other methods have been proposed for control of such systems and they can be summarized as follows:

1. **Open loop (feed-forward) control** is based on varying behavior of the nonlinear system under the action of predetermined external input. This approach is simple because it works without any measurements or sensors. This is especially important for the control of superfast processes.
2. **Linear and nonlinear (feed-back) control** deals with the possibilities of using the traditional approaches, and methods of automatic control over the problems of chaos control are discussed in numerous papers. The desired aim can be reached sometimes even by means of the simple proportional law of control and feedback (Genesio et al., 1993). The potentialities of the dynamic feedbacks can be better realized by using the observers (Ushio, 1999). Other methods of the modern theory of nonlinear control such as the theory of center manifold, sliding mode control (Ammour et al., 2009; Yang et al., 2009; Vinagre et al., 2006), the backstepping procedure, the reset control (Beker et al., 2004; Vinagre et al., 2007), and the H_∞-optimal design (Pan and Yin, 1997) can be used to solve the problems of stabilization for the given state.
3. **Adaptive control** assumes the possibility of applying the methods of adaptation to the control of chaotic processes, where the parameters of the controlled plant are unknown and the information about the model structure more often than not is incomplete. A number of the existing methods of adaptation such as the methods of gradient and speed gradient, least squares, and maximum likelihood can be used to develop algorithms of adaptive control and parametric identification. A controller is usually designed using the reference model or the methods of linearization by feedback (Ladaci and Charef, 2006; Vinagre et al., 2002).
4. **Linearization of the Poincaré map** method can be formulated by the following key ideas: (1) designing controller by the discrete system model based on linearization of the Poincaré map and (2) using the property of recurrence of the chaotic trajectories and applying the control action only at the instants when the trajectory returns to some neighborhood of the desired state or given orbit.
5. **Time-delayed feedback** method considers the problem of stabilizing an unstable periodic orbit of a nonlinear system by a simple feedback law with time delay.

Sensitivity to the parameter, especially to the delay time, is a disadvantage of the control law.

6. **Neural network-based control** deals with the ability of neural networks to control and predict behavior of nonlinear systems. The various structures of neural networks for control and prediction of the processes in nonlinear chaotic systems can be found in literature.
7. **Fuzzy control** uses a description of system indeterminacy in terms of fuzzy models, provides specific versions of the control algorithms, which consists of four blocks: knowledge base, fuzzification, inference and defuzzification (Calvo and Cartwright, 1998).

6.3 Examples: Feed-Back Control of Chaotic Systems

As already mentioned, the control of chaos means the control of unstable systems. The proposed controller has to reduce chaotic oscillations to the regular ones or eliminate them.

The control of chaos has also been studied and observed in many experiments (e.g., Ahmad et al., 2004; Lenz and Obradovic, 1997; Pan and Yin, 1997; Ushio, 1999; Tavazoei and Haeri, 2008; Wang et al., 2009, etc.). Especially, the control of well-known Chua's system by sampled data has been studied (Yang and Chua, 1998). The main motivation for the control of chaos via sampled data is well-developed digital control techniques.

6.3.1 Sampled-Data Control of Chua's Oscillator

In this brief study the practical results from sampled-data feedback control (case of the control strategy) of the fractional-order chaotic dynamical system are presented. The system was modeled by the state equation $\dot{\mathbf{x}} = \mathbf{f}(\mathbf{x})$, where $\mathbf{x} \in \mathrm{R}^n$ is state variable, $\mathbf{f}: \mathrm{R}^n \rightarrow \mathrm{R}^n$ is nonlinear function and $\mathbf{f}(0) = 0$.

The structure of control system with sampled data (Yang and Chua, 1998) is shown in Fig. 6.1. The state variables of the chaotic system are measured and the result is used to construct the output signal $\mathbf{y}(t) = D\mathbf{x}(t)$, where D is a constant matrix. The output $\mathbf{y}(t)$ is then sampled by sampling block to obtain $\mathbf{y}(k) = D\mathbf{x}(k)$ at the discrete moments kT, where $k = 0, 1, 2, \ldots$ and T is the sampling period. Then $D\mathbf{x}(k)$ is used by the controller to calculate the control signal $\mathbf{u}(k)$, which is fed back into chaotic system.

The controlled chaotic system is defined by the following relations (Yang and Chua, 1998):

$$\begin{aligned} \frac{d\mathbf{x}(t)}{dt} &= \mathbf{f}(\mathbf{x}(t)) + B\mathbf{u}(k), \quad t \in [kT, (k+1)T) \\ \mathbf{u}(k+1) &= C\mathbf{u}(k) + D\mathbf{x}(k), \quad k = 0, 1, 2 \ldots, \end{aligned} \tag{6.2}$$

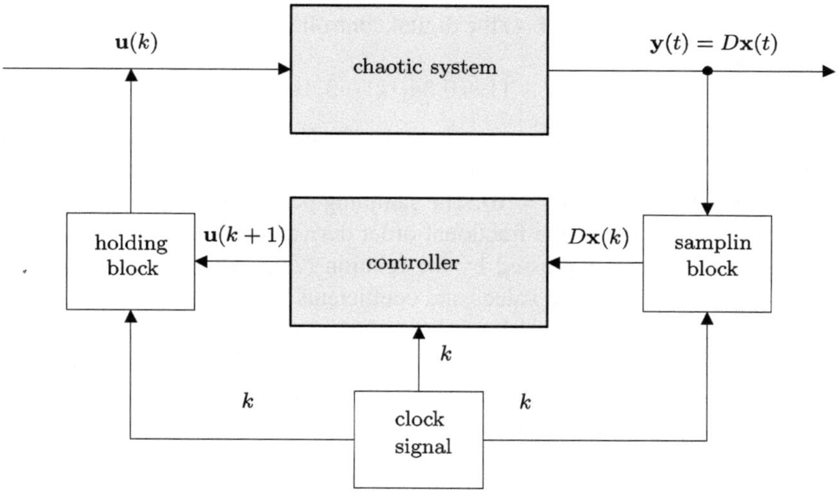

Fig. 6.1 Structure of the control system.

where $\mathbf{u} \in \mathrm{R}^m$, $B \in \mathrm{R}^n \times \mathrm{R}^m$, $C \in \mathrm{R}^m \times \mathrm{R}^m$, $D \in \mathrm{R}^m \times \mathrm{R}^n$ and $t \in \mathrm{R}_+$; $\mathbf{x}(k)$ is the sampled value of $\mathbf{x}(t)$ at $t = kT$. We suppose that $(0;0;0)$ is an equilibrium point of the system (6.2).

The approach described in this example is concentrating on the feedback control of the chaotic fractional-order Chua's system, where the total order of the system is 2.9. The controlled fractional-order Chua's system is defined by (Petráš, 2002):

$$\begin{aligned} \frac{dx_1(t)}{dt} &= \alpha\, {}_0D_t^{1-q}(x_2(t) - x_1(t) - f(x_1)) + u_1(t), \\ \frac{dx_2(t)}{dt} &= x_1(t) - x_2(t) + x_3(t) + u_2(t), \\ \frac{dx_3(t)}{dt} &= -\beta x_2(t) + u_3(t), \end{aligned} \tag{6.3}$$

where as a function $f(x_1)$ we consider a piecewise-linear nonlinearity (5.5). The linearized output $y(t)$ is given as $y(t) = x_1(t)/\alpha + x_3(t)/\beta$.

For numerical simulations the following parameters of the fractional Chua's system (6.3) are chosen:

$$\alpha = 10,\ \beta = \frac{100}{7},\ q = 0.9,\ m_0 = -1.27,\ m_1 = -0.68,$$

and the following parameters (experimentally found) of the controller:

$$B = \begin{pmatrix} 1\,0\,0 \\ 0\,0\,0 \\ 0\,0\,0 \end{pmatrix},\ C = \begin{pmatrix} 0.8\;0\;0 \\ 0\;\;0\;0 \\ 0\;\;0\;0 \end{pmatrix},\ D = \begin{pmatrix} -3.3\;0\;0 \\ 0\;\;\;0\;0 \\ 0\;\;\;0\;0 \end{pmatrix}. \tag{6.4}$$

Using the above parameters (6.4) the digital controller in state space feedback form is defined as

$$u_1(k+1) = 0.8u_1(k) - 3.3x_1(k), \tag{6.5}$$

for $k = 0, 1, 2, \ldots$ where $u_2(t) = 0$ and $u_3(t) = 0$. The initial conditions for Chua's circuit were $(x_1(0), x_2(0), x_3(0)) = (0.2, -0.1, -0.01)$ and the initial condition for the controller (6.5) was $u_1(0) = (0)$. The sampling period was $T = 0.01\,s$.

For the computation of the fractional-order derivative in Eqs. (6.3) a numerical approximation method expressed by the relation (2.53) was used. The length of memory was $L_m = 10$ (1000 values and coefficients from the history for sampling time $T = 0.01\,s$).

Fig. 6.2 shows the attractor of Chua's circuit (6.3) without control for the simulation time $100s$. Similar behavior was shown in the work (Hartley et al., 1995), where piecewise linear nonlinearity was replaced by cubic nonlinearity which yields very similar properties.

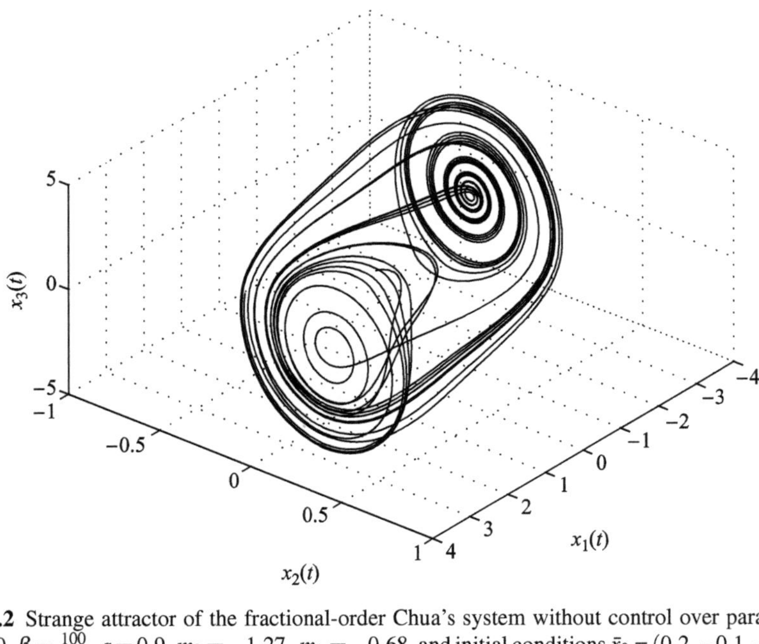

Fig. 6.2 Strange attractor of the fractional-order Chua's system without control over parameters $\alpha = 10$, $\beta = \frac{100}{7}$, $q = 0.9$, $m_0 = -1.27$, $m_1 = -0.68$, and initial conditions $\bar{x}_0 = (0.2, -0.1, -0.01)$.

In Fig. 6.3 is shown the controlled trajectory of state variables of the fractional-order Chua's system (6.3), which tends to origin asymptotically.

In Fig. 6.4 is shown the trajectory of the fractional-order Chua's system under control in state space for the simulation time $100s$.

In Fig. 6.5 is shown control signal from the digital controller (6.5). We can observe that signal tends to origin asymptotically similar to state variables behavior.

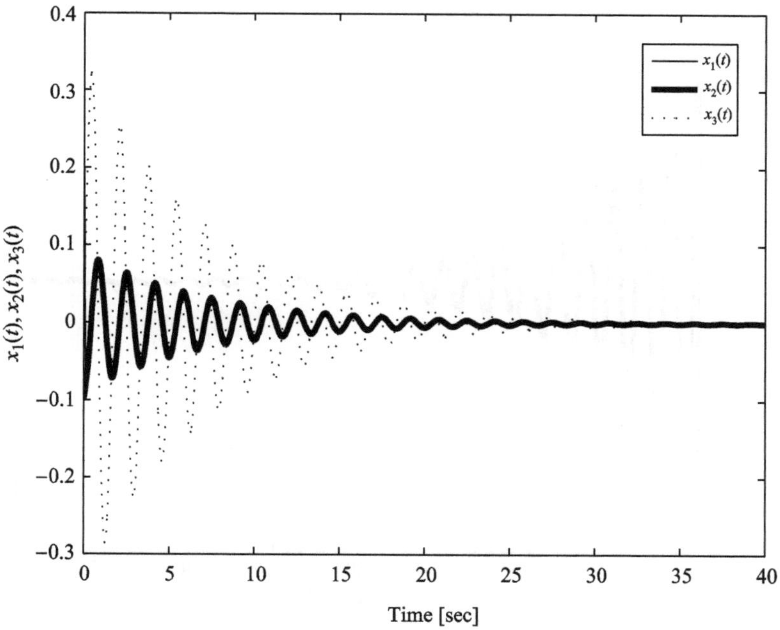

Fig. 6.3 Controlled state variables $x_1(t)$, $x_2(t)$, and $x_3(t)$.

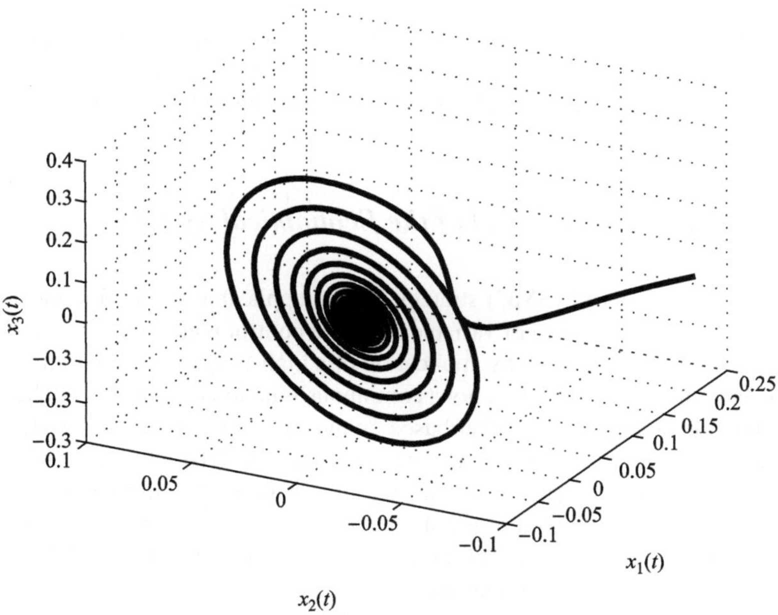

Fig. 6.4 The trajectory of the controlled fractional-order Chua's system in state space.

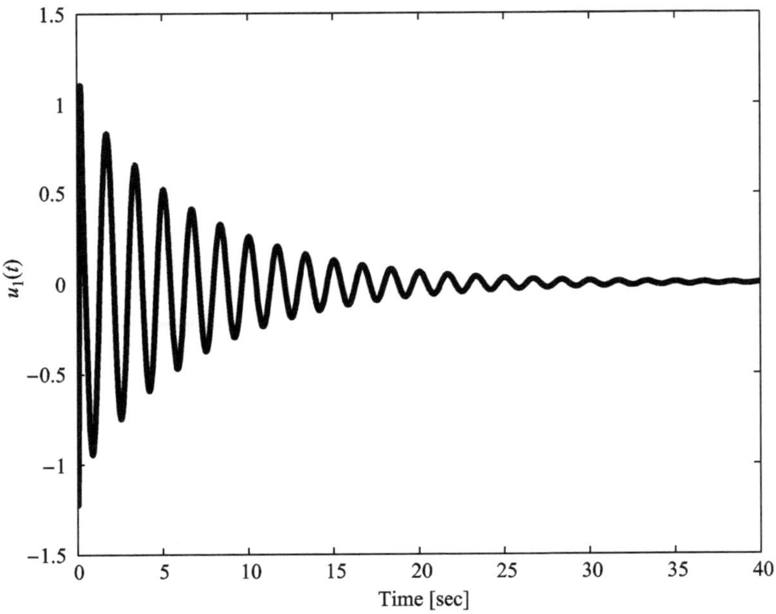

Fig. 6.5 Control signal $u_1(t)$.

A different type of the fractional-order chaotic system can be controlled by using this control strategy as well. It is a uniform approach to the control of chaos where the discrete proportional feedback controller is used (Jie and Lian, 2010).

6.3.2 Sliding Mode Control of the Economical System

A sliding model control (SMC) strategy is also applicable to the fractional-order chaotic systems. It is a form of variable structure control method that alters the dynamics of a nonlinear system by application of a high-frequency switching control. The state feedback control law is not a continuous function of time. It switches from one continuous structure to another based on the current position in the state space. Trajectories always move toward a switching condition. The motion of the system as it slides along these boundaries is called a sliding mode. The sliding mode control scheme involves: (i) selection of the sliding surface $\sigma(t)$ that represents a desirable system dynamic behavior, (ii) finding a switching control law that a sliding mode exists on every point of the sliding surface.

Consider the following general structure of the fractional-order nonlinear system under control

$$ {}_0D_t^q \mathbf{x}(t) = \mathbf{f}(\mathbf{x}(t)) + B\mathbf{u}(t), \tag{6.6}$$

where $\mathbf{u}(t) = [u_1(t)\, u_2(t) \cdots u_m(t)]^T$ is an m-dimensional input vector that will be used and the following control structure will be considered for state feedback:

$$\mathbf{u}(t) = u_{eq}(t) + u_{sw}(t), \tag{6.7}$$

where $u_{eq}(t)$ is the equivalent control and $u_{sw}(t)$ is the switching control of the system (6.6). A common task is to design a state feedback control law to stabilize the dynamical system (6.6) around the origin $\mathbf{x}(t) = [0,0,\ldots,0]^T$. In the sliding mode, the sliding surface and its derivative must satisfy $\sigma(t) = 0$ and $\dot{\sigma}(t) = 0$.

Let us use the controlled fractional-order financial system in the following form (Dadras and Momeni, 2010):

$$\begin{aligned}
{}_0D_t^{q_1} x_1(t) &= x_3(t) + (x_2(t) - a)x_1(t),\\
{}_0D_t^{q_2} x_2(t) &= 1 - bx_2(t) - x_1^2(t) + u(t),\\
{}_0D_t^{q_3} x_3(t) &= -x_1(t) - cx_3(t),
\end{aligned} \tag{6.8}$$

where a is the saving amount, b is the cost per investment, and c is the elasticity of demand of commercial market, $(a,b,c) \in \mathrm{R}^+$. The state variables $x_1(t)$, $x_2(t)$, and $x_3(t)$ are the interest rate, the investment demand, and the price index, respectively.

The proposed fractional sliding surface is defined as

$$\sigma(t) = \int_0^t (x_1^2(\tau) + Kx_2(\tau))d\tau + {}_0D_t^{q_2-1} x_2(t), \tag{6.9}$$

where K is a positive constant, in addition, $K \equiv K_{eq}$. The equivalent control $u_{eq}(t)$ is obtained by setting the derivative of sliding surface to zero and then solving the second equation of (6.8) for $u(t)$. We obtain

$${}_0D_t^{q_2} x_2(t) = -(x_1^2(t) + Kx_2(t))$$

and then we get the relation

$$\begin{aligned}
u_{eq}(t) &= {}_0D_t^{q_2} x_2(t) - 1 + bx_2(t) + x_1^2(t)\\
&= -x_1^2(t) + K_{eq}x_2(t) - 1 + bx_2(t) + x_1^2(t) = (b - K_{eq})x_2(t) - 1,
\end{aligned} \tag{6.10}$$

where K_{eq} is the constant of the controller.

The switching control $u_{sw}(t)$ law is chosen in order to satisfy the sliding condition

$$u_{sw}(t) = K_{sw}\mathrm{sign}(\sigma), \tag{6.11}$$

where

$$\mathrm{sign}(\sigma) = \begin{cases} +1, & \sigma > 0,\\ 0, & \sigma = 0,\\ -1, & \sigma < 0. \end{cases}$$

and K_{sw} is the gain of the controller $(K_{sw} < 0)$.

Finally, the total control law is defined as

$$u(t) = u_{eq}(t) + u_{sw}(t) = (b - K_{eq})x_2(t) - 1 + K_{sw}\text{sign}(\sigma). \tag{6.12}$$

We assume the following parameters of the chaotic system (6.8): $a = 1$, $b = 0.1$, $c = 1$, and the controller (6.12) parameters (experimentally found): $K_{eq} = 1.5$ and $K_{sw} = -3.5$. The controller will be applied at $t = 15\,s$. In the first case we use a commensurate order of derivatives $q_1 = q_2 = q_3 = 0.9$ and in the second case we use an incommensurate order of the derivatives $q_1 = 1.0$, $q_2 = 0.95$, and $q_3 = 0.99$ of the fractional-order chaotic system (6.8). The initial conditions for both cases are $(x_1(0), x_2(0), x_3(0)) = (2, -1, 1)$.

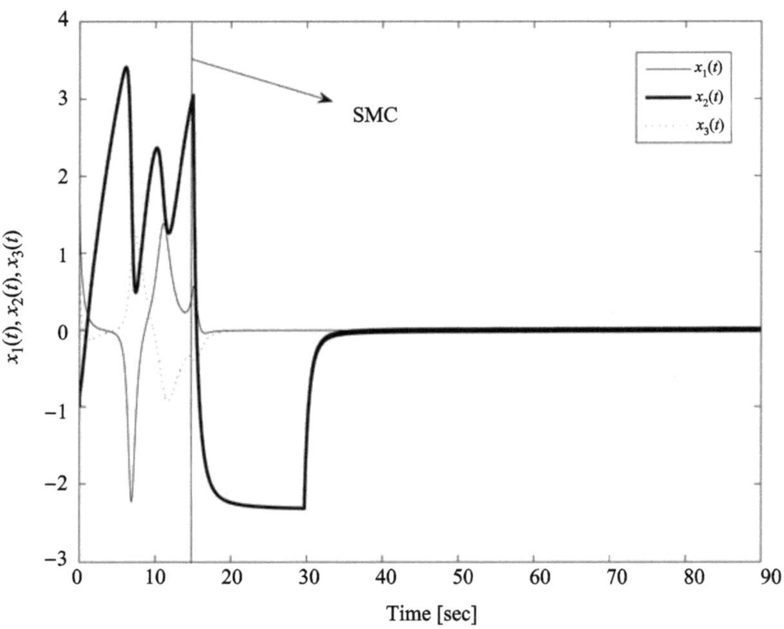

Fig. 6.6 Controlled state variables $x_1(t)$, $x_2(t)$, and $x_3(t)$ of commensurate fractional order financial system, where SMC was activated at $15\,s$.

In Fig. 6.6 are depicted the controlled state variables of the commensurate fractional-order financial systems (6.8) with the parameters: $a = 1$, $b = 0.1$, $c = 1$, orders $q_1 = q_2 = q_3 = 0.9$, controller (6.12) parameters: $K_{eq} = 1.5$ and $K_{sw} = -3.5$, initial conditions: $(x_1(0), x_2(0), x_3(0)) = (2, -1, 1)$ for simulation time $T_{sim} = 90\,s$ and time step $h = 0.005$.

In Fig. 6.7 is shown the control law of commensurate fractional-order financial system which drives the system states to the sliding surface. We can observe chattering in the sliding mode.

In Fig. 6.8 are depicted the controlled state variables of the incommensurate fractional-order financial systems (6.8) with the parameters: $a = 1$, $b = 0.1$, $c = 1$,

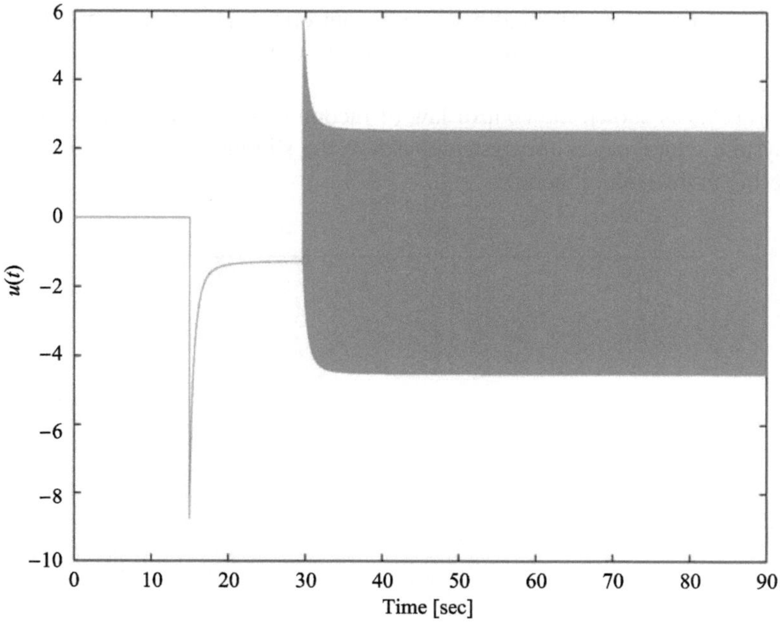

Fig. 6.7 Time response of control law $u(t)$ to commensurate fractional-order financial system, where SMC was activated at $15\,s$.

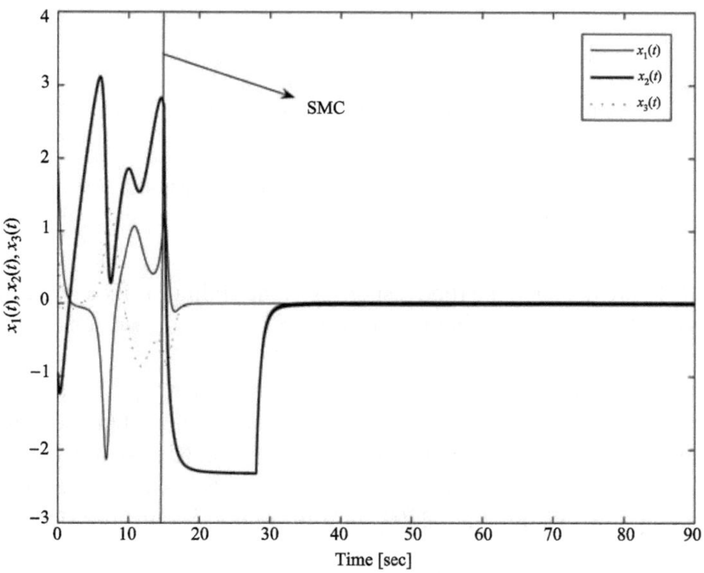

Fig. 6.8 Controlled state variables $x_1(t)$, $x_2(t)$, and $x_3(t)$ of incommensurate fractional order financial system, where SMC was activated at $15\,s$.

orders $q_1 = 1.0$, $q_2 = 0.95$, and $q_3 = 0.99$, controller (6.12) parameters: $K_{eq} = 1.5$ and $K_{sw} = -3.5$, initial conditions: $(x_1(0), x_2(0), x_3(0)) = (2, -1, 1)$ for simulation time $T_{sim} = 90\,s$ and time step $h = 0.005$.

In Fig. 6.9 is shown the control law of incommensurate fractional-order financial system which drives the system states to the sliding surface. We can observe chattering in the sliding mode.

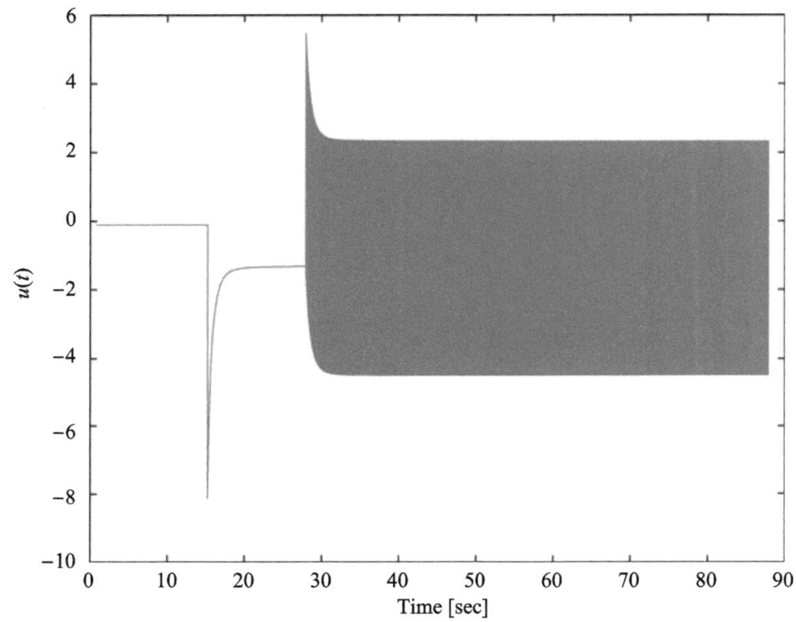

Fig. 6.9 Time response of control law $u(t)$ for incommensurate fractional-order financial system, where SMC was activated at $15\,s$.

In Fig. 6.10 are depicted the time responses of the sliding surface. We can observe that the controller kept the system states on the sliding surface for all subsequent time.

Performed simulations show that system responses after applying the control law (6.12) are satisfactory for both cases. The results confirm that obtained control strategy is efficient for controlling the fractional-order financial system (6.8).

To solve the chattering problem depicted in Fig. 6.7 and Fig. 6.9 we can use a new reaching law (Wang et al., 2009) or a fuzzy logic controller (Delavari et al., 2010).

The proposed control method is simple and control law is achieved to asymptotically stabilize the system if the controller is applied to the investment demand in order to control the whole economical system. This approach is applicable to different types of fractional-order chaotic systems as well (e.g., Wang et al., 2009; Yang et al., 2009, etc.).

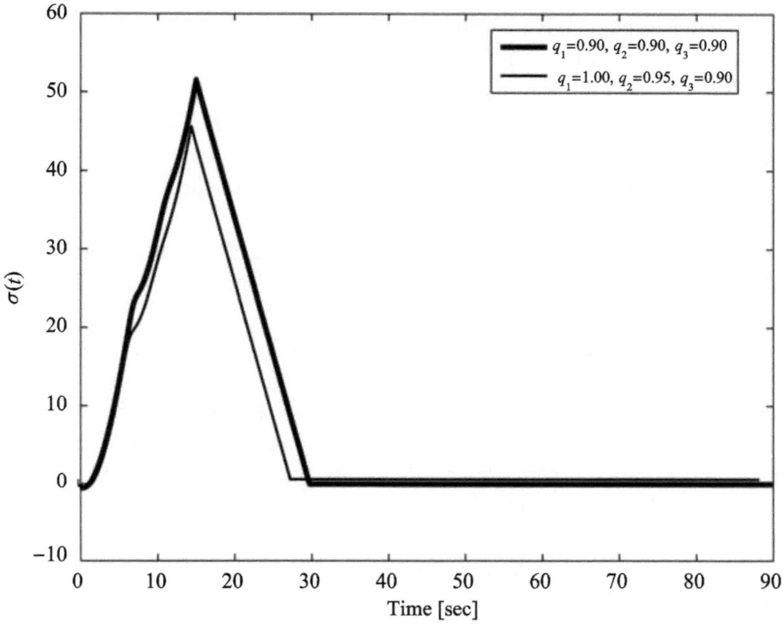

Fig. 6.10 Time response of sliding surface $\sigma(t)$ to commensurate and incommensurate fractional-order financial system, where SMC was activated at $15s$.

References

Ahmad W. M., El-Khazali R. and Al-Assaf Y., 2004, Stabilization of generalized fractional order chaotic systems using state feedback control, *Chaos, Solitons and Fractals*, **22**, 141–150.

Ammour A. S., Djennoune S. and Bettayeb M., 2009, A sliding mode control for linear fractional systems with input and state delays, *Commun. Nonlinear Sci. Numer. Simulat.*, **14**, 2310–2318.

Andrievskii B. R. and Fradkov A. L., 2003, Control of chaos: Methods and applications. I. Methods, *Automation and Remote Control*, **64**, 673–713.

Andrievskii B. R. and Fradkov A. L., 2004, Control of chaos: Methods and applications. II. Applications, *Automation and Remote Control*, **65**, 505–533.

Beker O., Hollot C. V., Chait Y. and Han H., 2004, Fundamental properties of reset control systems, *Automatica*, **40**, 905–915.

Calvo O. and Cartwright J. H. E., 1998, Fuzzy control of chaos, *Int. J. Bifurcat. Chaos. Appl. Sci. Eng.*, **8**, 1743–1747.

Caponetto R., Dongola G., Fortuna L. and Petráš I., 2010, *Fractional Order Systems: Modeling and Control Applications*, World Scientific, Singapore.

Chen Y. Q., Petráš I. and Xue D., 2009, Fractional order control - A tutorial, *Proc. of the American Control Conference, ACC 2009.*, June 10-12 2009, St. Louis, USA, 1397–1411.

Dadras S. and Momeni H. R., 2010, Control of a fractional-order economical system via sliding mode, *Physica A*, DOI: 10.1016/j.physa.2010.02.025.

Delavari H., Ghaderi R., Ranjbar A. and Momani S., 2010, Fuzzy fractional order sliding mode controller for nonlinear systems, *Commun. Nonlinear Sci. Numer. Simulat*, **15**, 963–978.

Deng W. H. and Li C. P., 2005, Chaos synchronization of the fractional Lü system, *Physica A*, **353**, 61–72.

Genesio R., Tesi A. and Villoresi F., 1993, A frequency approach for analyzing and controlling chaos in nonlinear circuits, *IEEE Trans. Circ. Syst. Fund. Theor. Appl.*, **40**, 819–828.

Hartley T. T., Lorenzo C. F. and Qammer H. K., 1995, Chaos on a fractional Chua's system, *IEEE Trans. Circ. Syst. Fund. Theor. Appl.*, **42**, 485–490.

Hosseinnia S. H., Ghaderi R., Ranjbar A., Abdous F. and Momani S., 2010, Control of chaos via fractional-order state feedback controller, *New Trends in Nanotechnology and Fractional Calculus Applications*, Springer, Part **V**, 511–519.

Jie Y. and Lian Q. D., 2010, The feedback control of fractional order unified chaotic system, *Chinese Phys. B*, **19**.

Ladaci S. and Charef A., 2006, On fractional adaptive control, *Nonlinear Dyn.*, **43**, 365–378.

Lenz H. and Obradovic D., 1997, Robust control of the chaotic Lorenz system, *Int. J. Bifurcat. Chaos. Appl. Sci. Eng.*, **7**, 2847–2854.

Li C. P., Deng W. H. and Xu D., 2006, Chaos synchronization of the Chua system with a fractional order, *Physica A*, **360**, 171–185.

Petráš I., 2002, Control of fractional-order Chua's system, *Journal of Electrical Engineering*, **53**, 219–222.

Pan S. and Yin F., 1997, Optimal control of chaos with synchronization, *Int. J. Bifurcat. Chaos. Appl. Sci. Eng.*, **7**, 2855–2860.

Pecora L. M. and Carroll T. L., 1990, Synchronization in chaotic systems, *Phys. Rev. Lett.*, **64**, 821–824.

Peng G., 2007, Synchronization of fractional order chaotic systems, *Physics Letters A*, **363**, 426–432.

Tavazoei M. S. and Haeri M., 2008, Chaos control via a simple fractional-order controller, *Physics Letters A*, **372**, 798–807.

Tavazoei M. S., Haeri, M., Bolouki S. and Siami M., 2009, Using fractional-order integrator to control chaos in single-input chaotic systems, *Nonlinear Dyn.*, **55**, 179-190.

Ushio T., 1999, Synthesis of synchronized chaotic systems based on observers, *Int. J. Bifurcat. Chaos. Appl. Sci. Eng.*, **9**, 541–546.

Vinagre B. M., Petráš I., Podlubny I. and Chen Y. Q., 2002, Using fractional order adjustment rules and fractional order reference models in model-reference adaptive control, *Nonlinear Dyn.*, **29**, 269–279.

Vinagre B. M. and Calderón A. J., 2006, On fractional sliding mode control, *Proc. of the 7th Portuguese Conference on Automatic Control – CONTROLO 2006*, Instituto Superior Técnico, September 11–13, Lisboa, Portugal.

Vinagre B. M., Monje C. A. and Tejado I., 2007, Reset and fractional integrators in control applications, *Proc. of the Int. Carpathian Control Conf.*, High Tatras, Slovakia, May 24–27, 754–757.

Wang G. and Chen Y., 2008, Control of fractional order chaotic systems, *Proc. of the Second Int. Conf. on Genetic and Evolutionary Computing*, 25–26 Sept. 2008, Hubei, 535–538.

Wang H., Hana Z., Xiea Q. and Zhang W., 2009, Sliding mode control for chaotic systems based on LMI, *Commun. Nonlinear Sci. Numer. Simulat.*, **14**, 1410–1417.

Yang Y. S., Lin J. S., Liao T. L. and Yan J. J., 2009, Sliding mode control design for fractional chaotic systems, *Proc. of the 2009 IEEE Int. Conference on Networking, Sensing and Control*, Okayama, Japan, March 26–29, 2009, 539–542.

Yang T. and Chua L. O., 1998, Control of chaos using sampled-data feedback control, *Int. J. Bifurcat. Chaos. Appl. Sci. Eng.*, **8**, 2433–2438.

Zhu H., Zhou S. and He Z., 2009, Chaos synchronization of the fractional-order Chen's system, *Chaos, Solitons and Fractals*, **41**, 2733–2740.

Chapter 7
Conclusion

In this book the fractional-order nonlinear systems and methods for their numerical simulation and stability analysis are presented. By illustrative examples we have shown chaotic behaviour of such systems and studied their dynamics. We presented the examples of electrical, mechanical, hydrodynamical, chemical, biological, economical, and the other chaotic systems. We studied only the state trajectories (attractors) and we avoided bifurcation analysis and Poincaré maps.

Some authors consider the attractors of chaotic systems a numerical error (Yao, 2010). In fact, deterministic chaos exists if the Lyapunov exponent of the system is positive (Parker and Chua, 1989). We also presented the so-called instability measure as a condition to determine chaos in fractional-order systems. Computation of strange attractors in fractional-order nonlinear systems is very important and therefore we have to find appropriate approximation methods. Utilization of methods in the form of rational polynomial leads to high-order systems. In this case we must consider different initial conditions and large numerical errors which are amplified by the systems constants and approximation polynomial constants. We recommend using a method in the form of FIR filter with a large number of coefficients because it works more accurately and numerical errors are much smaller than those of the methods in the form of IIR filter (Vinagre et al., 2003). However, the time of computation is longer because of the number of coefficients.

In this book we also mention a total order of fractional-order systems. The system order in such case is equal to the sum of particular fractional orders of differential equations. The conclusion of this work confirms the conclusions of the works (Arena et al., 2000; Hartley et al., 1995; Podlubny, 1999) that there is a need to refine the notion of the order of a system which cannot be considered only by the total order of differentiation. For fractional-order differential equations the number of terms in equations and the number of equations are more important than the order of differentiation.

We have considered examples of chaotic fractional-order systems which exhibit chaotic behavior, with total order less than three except Duffing's, Van der Pol's oscillators and Lotka-Volterra system with total order less than two, and memristor based Chua's oscillator with hyperchaos and total order less than four. We have shown chaotic systems with several types of nonlinearities as for example piecewise-linear nonlinearity, cross product, square and cubic power and so on. For these fractional-order chaotic systems we have made:

- mathematical description,
- stability investigation,
- numerical solution,
- Matlab routines for simulation,
- Matlab/Simulink models (for two systems).

The Matlab functions have been created for all described chaotic systems and they are listed in Appendix A. The Simulink models have been created only for two systems, namely, fractional-order Chua's and Volta's systems as a general guide for such kind of system simulation.

There are a large number of fractional-order chaotic systems that are not described in this book. This number rapidly grows (e.g., Caponetto et al., 2010; Hilfer, 2000; West et al., 2002; Zaslavsky, 2005, etc.). To have a closer picture we refer to several additional references but we have to note that this list is not complete. For illustration we can mention additional well-known fractional-order chaotic systems, e.g., delayed fractional-order chaotic systems (Deng et al., 2007; Guo, 2006), hyperchaotic systems (Ahmad, 2005b; Deng et al., 2009; Matouk, 2009), fractional-order HIV model (Ye and Ding, 2009), fractional-order multi-scroll attractors system (Ahmad, 2005a; Deng and Lu, 2007), fractional-order 3-D quadratic autonomous system with 4-wing attractor (Wang et al., 2010), fractional-order Sprott's electronic oscillator and mechanical "jerk" model (Ahmad and Sprott, 2003), fractional neuron network system (Zhou et al., 2008), etc. In addition, we note that there are various modifications of Chen's system as, for example, hybrid Lorenz-Chen system (Lian at al., 2007) or Chen-Lee system (Tam and Tou, 2008). There are also many works which report possible electronic implemenation of such type chaotic systems (e.g., Li et al., 2009; Tavazoei et al., 2008, etc.) and its utilization, for instance, in a chaotic secure communication scheme (Kiani-B et al., 2009).

Various numerical methods may also be used in chaotic attractor computations. In addition to the proposed algorithm based on Grünwald-Letnikov definition of the fractional derivative, a modified matrix approach (Podlubny, 2000; Podlubny et al., 2009) or the Adams-Bashforth-Moulton type predictor-corrector scheme (Deng, 2007a,b; Diethelm et al., 2005; Ford and Simpson, 2001) can be successfully used. The frequency-based methods are not sufficient because as shown in (Tavazoei and Haeri, 2007a,b, 2008), false chaos can be observed in systems, which are not chaotic. It is influenced by approximation error.

Some remarks on chaos control have been noted as well. We mentioned several control strategies and synchronization techniques and by illustrative examples presented control of chaos via feedback methods. Two such methods are described: (i) digital state-space proportional feedback controller, and (ii) sliding mode controller.

Finally, for detecting chaotic behavior in the system an instability measure can be used. Computation of the Lyapunov exponents is sometimes impossible and instead of these exponents, the instability measure is a sufficient condition for detecting chaos in the fractional-order chaotic systems.

As has been demonstrated, the idea of fractional calculus requires one to reconsider dynamic system concepts that are often taken for granted. So by changing the order of a system from integer to real, we also move from a three-dimensional system to infinite dimension. A lot of tasks have been opened, namely, stability analysis of uncertain nonlinear fractional-order systems, conditions to determine chaos, control strategies and so on. They should be considered in further work.

Besides mentioned one we also have to note a problem related to an identification of the fractional-order chaotic system parameters (Al-Assaf et al., 2004). It is a difficult task because any change in the system fractional derivative orders or system coefficients generates completely different time response. It is necessary to find an effective identification technique in order to obtain dynamical models that represent the given measured chaotic data in finite time. It could bring a lot of possible applications such as, for example, modelling of the macroeconomic performance of the countries (Petráš and Podlubny, 2007), or many other interesting phenomena with chaotic nature.

References

Ahmad W. M. and Sprott J. C., 2003, Chaos in fractional-order autonomous nonlinear systems, *Chaos, Solitons and Fractals*, **16**, 339–351.

Ahmad W. M., 2005a, Generation and control of multi-scroll chaotic attractors in fractional order systems, *Chaos, Solitons and Fractals*, **25**, 727–735.

Ahmad W. M., 2005b, Hyperchaos in fractional order nonlinear systems, *Chaos, Solitons and Fractals*, **26**, 1459–1465.

Arena P., Caponetto R., Fortuna L. and Porto D., 2000, *Nonlinear Noninteger Order Circuits and Systems – An Introduction*, World Scientific, Singapore.

Al-Assaf Y., El-Khazali R. and Ahmad W., 2004, Identification of fractional chaotic system parameters, *Chaos, Solitons and Fractals*, **22**, 897–905.

Caponetto R., Dongola G., Fortuna L. and Petráš I., 2010, *Fractional Order Systems: Modeling and Control Applications*, World Scientific, Singapore.

Deng H., Li T., Wang Q. and Li H., 2009, A fractional-order hyperchaotic system and its synchronization, *Chaos, Solitons and Fractals*, **41**, 962–969.

Deng W., Li Ch. and Lu J., 2007, Stability analysis of linear fractional differential system with multiple time delays, *Nonlinear Dynamics*, **48**, 409–416.

Deng W., 2007a, Short memory principle and a predictor-corrector approach for fractional differentional equations, *Journal of Computational and Applied Mathematics*, **206**, 174–188.

Deng W., 2007b, Numerical algorithm for the time fractional Fokker-Planck equation, *Journal of Computational Physics*, **227**, 1510–1522.

Deng W. and Lu J., 2007, Generating multi-directional multi-scroll chaotic attractors via a fractional differential hysteresis system, *Physics Letters A*, **369**, 438–443.

Diethelm K., Ford N. J., Freed A. D. and Luchko Yu., 2005, Algorithms for the fractional calculus: A selection of numerical methods, *Comput. Methods Appl. Mech. Engrg.*, **194**, 743–773.

Ford N. and Simpson A., 2001, The numerical solution of fractional differential equations: speed versus accuracy, *Num. Anal. Report 385*, Manchester Centre for Computational Mathematics.

Guo L. J., 2006, Chaotic dynamics of the fractional-order Ikeda delay system and its synchronization, *Chinese Physics*, **15**, 301–305.

Hartley T. T., Lorenzo C. F. and Qammer H. K., 1995, Chaos on a fractional Chua's system, *IEEE Trans. Circ. Syst. Fund. Theor. Appl.*, **42**, 485–490.

Hilfer R., 2000, *Application of fractional calculus in physics*, World Scientific, Singapore.

Kiani-B A., Fallahi K., Pariz N. and Leung H., 2009, A chaotic secure communication scheme using fractional chaotic systems based on an extended fractional Kalman filter, *Communications in Nonlinear Science and Numerical Simulation*, **14**, 863–879.

Li X. F., Chlouverakis K. E. and Xu D. L., 2009, Nonlinear dynamics and circuit realization of a new chaotic flow: A variant of Lorenz, Chen and Lü, *Nonlinear Analysis: Real World Applications*, **10**, 2357–2368.

Lian Q. D., Qiao W. and Hong G., 2007, Chaotic attractor transforming control of hybrid Lorenz-Chen system, *Chinese Phys. B*, **17**.

Matouk A. E., 2009, Stability conditions, hyperchaos and control in a novel fractional order hyperchaotic system, *Physics Letters A*, **373**, 2166–2173.

Parker T. S. and Chua L. O., 1989, *Practical Numerical Algorithm for Chaotic Systems*, Springer, New York.

Petráš I. and Podlubny I., 2007, State space description of national economies: the V4 countries, *Computational Statistics & Data Analysis*, **52**, 1223–1233.

Podlubny I., 1999, *Fractional Differential Equations*, Academic Press, San Diego.

Podlubny I., 2000, Matrix approach to discrete fractional calculus, *Fractional Calculus and Applied Analysis*, **3**, 359–386.

Podlubny I., Chechkin A., Škovránek T., Chen Y. Q. and Vinagre B. M. J., 2009, Matrix approach to discrete fractional calculus II: Partial fractional differential equations, *Journal of Computational Physics*, **228**, 3137–3153.

Tam L. M. and Tou W. M. S., 2008, Parametric study of the fractional-order Chen-Lee system, *Chaos, Solitons and Fractals*, **37**, 817–826.

Tavazoei M. S. and Haeri M., 2007a, Unreliability of frequency-domain approximation in recognising chaos in fractional-order systems, *IET Signal Proc.*, **1**, 171–181.

Tavazoei M. S. and Haeri, M., 2007b, A necessary condition for double scroll attractor existence in fractional-order systems, *Physics Letters A*, **367**, 102–113.

Tavazoei M. S., Haeri M., Jafari S., Bolouki S. and Siami M., 2008, Some Applications of Fractional Calculus in Suppression of Chaotic Oscillations, *IEEE Trans. Ind. Electron.*, **55**, 4094–4101.

Tavazoei M. S. and Haeri M., 2008, Limitations of frequency domain approximation for detecting chaos in fractional order systems, *Nonlinear Analysis*, **69**, 1299–1320.

Vinagre B. M., Chen Y. Q. and Petráš I., 2003, Two direct Tustin discretization methods for fractional-order differentiator/integrator, *J. Franklin Inst.*, **340**, 349–362.

Wang Z., Sun Y., Qi G. and van Wyk B. J., 2010, The effects of fractional order on a 3-D quadratic autonomous system with four-wing attractor, *Nonlinear Dyn.*, DOI: 10.1007/s11071-010-9705-7.

West B. J., Bologna M. and Grigolini P., 2002, *Physics of Fractal Operators*, Springer, New York.

Yao L. S., 2010, Computed chaos or numerical errors, *Nonlinear Analysis: Modelling and Control*, **15**, 109–126.

Ye H. and Ding Y., 2009, Nonlinear Dynamics and Chaos in a Fractional-Order HIV Model, *Mathematical Problems in Engineering*, Article ID **378614**.

Zaslavsky G. M., 2005, *Hamiltonian Chaos and Fractional Dynamics*, Oxford University Press, Oxford.

Zhou S., Li H. and Zhu Z., 2008, Chaos control and synchronization in a fractional neuron network system, *Chaos, Solitons and Fractals*, **36**, 973–984.

Appendix A
A List of Matlab Functions

A list of Matlab functions created for simulation of fractional-order chaotic systems:

- `function [T, Y] = FOChuaNR(parameters, orders, TSim, Y0)`
- `function [T, Y] = FOChuaM(parameters, orders, TSim, Y0)`
- `function [T, Y] = FOvanDerPol(parameters, orders, TSim, Y0)`
- `function [T, Y] = FODuffing(parameters, orders, TSim, Y0)`
- `function [T, Y] = FOGenTesi(parameters, orders, TSim, Y0)`
- `function [T, Y] = FOArneodo(parameters, orders, TSim, Y0)`
- `function [T, Y] = FOLorenz(parameters, orders, TSim, Y0)`
- `function [T, Y] = FORossler(parameters, orders, TSim, Y0)`
- `function [T, Y] = FOLu(parameters, orders, TSim, Y0)`
- `function [T, Y] = FOChen(parameters, orders, TSim, Y0)`
- `function [T, Y] = FOLotkaVolterra(parameters, orders, TSim, Y0)`
- `function [T, Y] = FONewLeipnik(parameters, orders, TSim, Y0)`
- `function [T, Y] = FOVolta(parameters, orders, TSim, Y0)`
- `function [T, Y] = FO3CNN(parameters, orders, TSim, Y0)`
- `function [T, Y] = FOLiu(parameters, orders, TSim, Y0)`
- `function [T, Y] = FOFinanc(parameters, orders, TSim, Y0)`

In all the above functions, variable `parameters` are for parameters of each chaotic system, variable `orders` are for real orders of the fractional derivatives in the chaotic system equations, variable *TSim* is the simulation time, and `Y0` are initial conditions. Returning parameters are: `T`-vector of the computional time for the $0 : h : TSim$, where h is the time step and `Y`-vector of the numerical solution for state variables $(Y(1), Y(2), \ldots)$, respectively.

Supporting Matlab function `memo()` is the so-called memory term used for calculation of the fractional derivative.

The full codes with demo and full description of each function are downloadable as a Matlab toolbox (Fractional Order Chaotic Systems) from FileExchange at the MathWorks, Inc. (http://www.mathworks.com/matlabcentral/fileexchange/27336/).

For illustration is presented an example of using the Matlab function created for the fractional-order Volta system (`FOVolta()`). By typing command

```
>>help FOVolta
```

we ge the following description of this function:

```
% Numerical Solution of the Fractional-Order Volta's System
%
%   D^q1 x(t) = -x(t) - a y(t) - z(t)y(t)
%   D^q2 y(t) = -y(t) - b x(t) - x(t)z(t)
%   D^q3 z(t) = c z(t) + x(t)y(t) + 1
%
% function [T, Y] = FOVolta( parameters, orders, TSim, Y0 )
%
% Input:     parameters - model parameters [a, b, c]
%            orders - derivatives orders [q1, q2, q3]
%            TSim - simulation time (0 - TSim) in sec
%            Y0 - initial conditions [Y0(1), Y0(2), Y0(3)]
%
% Output:    T - simulation time (0 : Tstep : TSim)
%            Y - solution of the system (x=Y(1),y=Y(2),z=Y(3))
%
% Author:   (c) Ivo Petras (ivo.petras@tuke.sk), 2010.
%
```

For instance, if we consider the following parameters of the fractional-order Volta system: $a = 19, b = 11, c = 0.73$ and the orders $q_1 = q_2 = q_3 = 0.98$, for the simulation time $T_{sim} = 20s$ and the initial conditions $(x(0), y(0), z(0)) = (8, 2, 1)$, it is necessary to call the following command in the Matlab environment:

```
>>[t, y] = FOVolta([19 11 0.73], [0.98 0.98 0.98], 20, [8 2 1]).
```

It is necessary to have a supporting function `memo()` in the same folder. The returning vector y consists of the numerical solution for the system for each particular state variable (x, y, z) in accordance to the vector of time variable `t`. Time step is set explicitly in each function to the default value $h = 0.005$. The results can be interpreted as follows:

- plotting the solution of state variable $x(t)$ for time $0 - 20s$ in black color:

```
>>plot(t, y(:,1),'k');
>>xlabel('Time [sec]');
>>ylabel('x(t)');
```

- plotting the solution of state variable $y(t)$ for time $0-20\,s$ in black color:

```
>>plot(t, y(:,2),'k');
>>xlabel('Time [sec]');
>>ylabel('y(t)');
```

- plotting the solution of state variable $z(t)$ for time $0-20\,s$ in black color:

```
>>plot(t, y(:,3),'k');
>>xlabel('Time [sec]');
>>ylabel('z(t)');
```

- plotting the trajectory of state variables $x(t), y(t), z(t)$ in state space in black color:

```
>>plot3(y(:,1), y(:,2), y(:,3), 'k');
>>xlabel('x(t)');
>>ylabel('y(t)');
>>zlabel('z(t)');
```

- plotting the trajectory of state variables $x(t), y(t)$ in state plane in black color:

```
>>plot(y(:,1), y(:,2) 'k');
>>xlabel('x(t)');
>>ylabel('y(t)');
```

- plotting the trajectory of state variables $x(t), z(t)$ in state plane in black color:

```
>>plot(y(:,1), y(:,3), 'k');
>>xlabel('x(t)');
>>ylabel('z(t)');
```

- plotting the trajectory of state variables $y(t), z(t)$ in state plane in black color:

```
>>plot(y(:,2), y(:,3), 'k');
>>xlabel('y(t)');
>>ylabel('z(t)').
```

If the grid in figures is necessary, the command `grid` is used as well. The above syntax is useful for all created functions and it has been used for simulations performed and described in this book. It is important to note that other Matlab functions are very helpful. For example, function `solve()` is useful for equilibrium calculation, function `eig()` returns eigenvalues, etc.

Appendix B
Laplace and Inverse Laplace Transforms

Table B.1 A list of Laplace and inverse Laplace transforms related to fractional-order calculus

F(s)	**f(t)**
$\frac{1}{\sqrt{s}}$	$\frac{1}{\sqrt{\pi t}}$
$\frac{1}{s\sqrt{s}}$	$2\sqrt{\frac{t}{\pi}}$
$\frac{1}{s^n\sqrt{s}},\ (n=1,2,\ldots)$	$\frac{2^n t^{n-(1/2)}}{1\cdot 3\cdot 5\cdots(2n-1)\sqrt{\pi}}$
$\frac{s}{(s-a)^{\frac{3}{2}}}$	$\frac{1}{\sqrt{\pi t}}e^{at}(1+2at)$
$\sqrt{s-a}-\sqrt{s-b}$	$\frac{1}{2\sqrt{\pi t^3}}(e^{bt}-e^{at})$
$\frac{1}{\sqrt{s}+a}$	$\frac{1}{\sqrt{\pi t}}-ae^{a^2t}\mathrm{erfc}(a\sqrt{t})$
$\frac{\sqrt{s}}{s-a^2}$	$\frac{1}{\sqrt{\pi t}}+ae^{a^2t}\mathrm{erfc}(a\sqrt{t})$
$\frac{\sqrt{s}}{s+a^2}$	$\frac{1}{\sqrt{\pi t}}-\frac{2a}{\sqrt{\pi}}e^{-a^2t}\int_0^{a\sqrt{t}}e^{\tau^2}\,\mathrm{d}\tau$
$\frac{1}{\sqrt{s}(s-a^2)}$	$\frac{1}{a}e^{a^2t}\mathrm{erfc}(a\sqrt{t})$
$\frac{1}{\sqrt{s}(s+a^2)}$	$\frac{2}{a\sqrt{\pi}}e^{-a^2t}\int_0^{a\sqrt{t}}e^{\tau^2}\,\mathrm{d}\tau$

Continued

F(s)	f(t)
$\frac{b^2-a^2}{(s-a^2)(\sqrt{s}+b)}$	$e^{a^2t}[b-a\mathrm{erfc}(a\sqrt{t})]-be^{b^2t}\mathrm{erfc}(b\sqrt{t})$
$\frac{1}{\sqrt{s}(\sqrt{s}+a)}$	$e^{a^2t}\mathrm{erfc}(a\sqrt{t})$
$\frac{1}{\sqrt{s+b}(s+a)}$	$\frac{1}{\sqrt{b-a}}e^{-at}\mathrm{erfc}(\sqrt{b-a}\sqrt{t})$
$\frac{b^2-a^2}{\sqrt{s}(s-a^2)(\sqrt{s}+b)}$	$e^{a^2t}\left[\frac{b}{a}\mathrm{erf}(a\sqrt{t})-1\right]+e^{b^2t}\mathrm{erfc}(b\sqrt{t})$
$\frac{(1-s)^n}{s^{n+(1/2)}}$	$\frac{n!}{(2n)!\sqrt{\pi t}}H_{2n}(\sqrt{t})$, where $H_n(x)=e^{x^2}\frac{\mathrm{d}^n}{\mathrm{d}x^n}(e^{-x^2})$ is Hermite polynomial.
$\frac{(1-s)^n}{s^{n+(3/2)}}$	$-\frac{n!}{(2n+1)!\sqrt{\pi}}H_{2n+1}(\sqrt{t})$
$\frac{\sqrt{s+2a}-\sqrt{s}}{\sqrt{s}}$	$ae^{-at}[I_1(at)+I_0(at)]$, where $I_n(x)=j^{-n}J_n(jt)$, J_n is Bessel function.
$\frac{1}{\sqrt{s+a}\sqrt{s+b}}$	$e^{-\frac{1}{2}(a+b)t}I_0\left(\frac{a-b}{2}t\right)$
$\frac{\Gamma(k)}{(s+a)^k(s+b)^k}, k\geq 0$	$\sqrt{\pi}\left(\frac{t}{a-b}\right)^{k-(1/2)}e^{-\frac{1}{2}(a+b)t}I_{k-(1/2)}\left(\frac{a-b}{2}t\right)$
$\frac{1}{(s+a)^{1/2}(s+b)^{3/2}}$	$te^{-\frac{1}{2}(a+b)t}\left[I_0\left(\frac{a-b}{2}t\right)+I_1\left(\frac{a-b}{2}t\right)\right]$
$\frac{\sqrt{s+2a}-\sqrt{s}}{\sqrt{s+2a}+\sqrt{s}}$	$\frac{1}{t}e^{-at}I_1(at)$
$\frac{(a-b)^k}{(\sqrt{s+a}+\sqrt{s+b})^{2k}}, k>0$	$\frac{k}{t}e^{-\frac{1}{2}(a+b)t}I_k\left(\frac{a-b}{2}t\right)$
$\frac{1}{\sqrt{s}\sqrt{s+a}(\sqrt{s+a}+\sqrt{s})^{2\nu}}, k>0$	$\frac{1}{a^\nu}e^{-\frac{1}{2}at}I_\nu\left(\frac{a}{2}t\right)$
$\frac{1}{\sqrt{s^2+a^2}}$	$J_0(at)$
$\frac{1}{\sqrt{s^2-a^2}}$	$I_0(at)$, modified Bessel function of the first kind, zeroth order.

Continued

F(s)	f(t)
$\dfrac{(\sqrt{s^2+a^2}-s)^\nu}{\sqrt{s^2+a^2}}, \nu > -1$	$a^\nu J_\nu(at)$
$\dfrac{1}{(\sqrt{s^2+a^2})^k}, k > 0$	$\dfrac{\sqrt{\pi}}{\Gamma(k)}\left(\dfrac{t}{2a}\right)^{k-(1/2)} J_{k-(1/2)}(at)$
$(\sqrt{s^2+a^2}-s)^k, k > 0$	$\dfrac{ka^k}{t} J_k(at)$
$\dfrac{(\sqrt{s^2-a^2}+s)^\nu}{\sqrt{s^2-a^2}}, \nu > -1$	$a^\nu I_\nu(at)$
$\dfrac{1}{(s^2-a^2)^k}, k > 0$	$\dfrac{\sqrt{\pi}}{\Gamma(k)}(\dfrac{t}{2a})^{k-(1/2)} I_{k-(1/2)}(at)$
$\dfrac{1}{s\sqrt{s+1}}$	$\mathrm{erfc}(\sqrt{t}) \quad \left(\mathrm{erfc}(x) = \dfrac{2}{\sqrt{\pi}}\int_0^x e^{-\tau^2}\mathrm{d}\tau = 1-\mathrm{erfc}(x)\right)$
$\dfrac{1}{s+\sqrt{s^2+a^2}}$	$\dfrac{J_1(at)}{at}$
$\dfrac{1}{(s+\sqrt{s^2+a^2})^N}$, where $N \in \mathbf{N}^+$	$\dfrac{NJ_N(at)}{a^N t}$
$\dfrac{1}{\sqrt{s^2+a^2}(s+\sqrt{s^2+a^2})}$	$\dfrac{J_1(at)}{a}$
$\dfrac{1}{\sqrt{s^2+a^2}(s+\sqrt{s^2+a^2})^N}$	$\dfrac{J_N(at)}{a^N}$
$\dfrac{k}{s^2+k^2}\coth\dfrac{\pi s}{2k}$	$\|\sin kt\|$
$\dfrac{1}{s}e^{-k/s}$	$J_0(2\sqrt{kt})$
$\dfrac{1}{\sqrt{s}}e^{-k/s}$	$\dfrac{1}{\sqrt{\pi t}}\cos 2\sqrt{kt}$
$\dfrac{1}{\sqrt{s}}e^{k/s}$	$\dfrac{1}{\sqrt{\pi t}}\cosh 2\sqrt{kt}$
$\dfrac{1}{s\sqrt{s}}e^{-k/s}$	$\dfrac{1}{\sqrt{\pi k}}\sin 2\sqrt{kt}$
$\dfrac{1}{s\sqrt{s}}e^{k/s}$	$\dfrac{1}{\sqrt{\pi k}}\sinh 2\sqrt{kt}$

Continued

F(s)	**f(t)**
$\frac{1}{s^{\nu}}e^{-k/s}, \nu > 0$	$\left(\frac{t}{k}\right)^{(\nu-1)/2} J_{\nu-1}(2\sqrt{kt})$
$\frac{1}{s^{\nu}}e^{k/s}, \nu > 0$	$\left(\frac{t}{k}\right)^{(\nu-1)/2} I_{\nu-1}(2\sqrt{kt})$
$e^{-k\sqrt{s}}, k > 0$	$\frac{k}{2\sqrt{\pi t^3}}e^{-\frac{k^2}{4t}}$
$\frac{1}{s}e^{-k\sqrt{s}}, k \geq 0$	$\operatorname{erfc}\left(\frac{k}{2\sqrt{t}}\right)$
$\frac{1}{\sqrt{s}}e^{-k\sqrt{s}}, k \geq 0$	$\frac{1}{\sqrt{\pi t}}e^{-\frac{k^2}{4t}}$
$\frac{1}{s\sqrt{s}}e^{-k\sqrt{s}}, k \geq 0$	$2\sqrt{\frac{t}{\pi}}e^{-\frac{k^2}{4t}} - k\operatorname{erfc}\left(\frac{k}{2\sqrt{t}}\right)$
$\frac{ae^{-k\sqrt{s}}}{s(a+\sqrt{s})}, k \geq 0$	$-e^{ak}e^{a^2t}\operatorname{erfc}\left(a\sqrt{t}+\frac{k}{2\sqrt{t}}\right)+\operatorname{erfc}\left(\frac{k}{2\sqrt{t}}\right)$
$\frac{e^{-k\sqrt{s}}}{\sqrt{s}(a+\sqrt{s})}, k \geq 0$	$e^{ak}e^{a^2t}\operatorname{erfc}\left(a\sqrt{t}+\frac{k}{2\sqrt{t}}\right)$
$\log\frac{s-a}{s-b}$	$\frac{1}{t}(e^{bt}-e^{at})$
$\log\frac{s^2+a^2}{s^2}$	$\frac{2}{t}(1-\cos at)$
$\log\frac{s^2-a^2}{s^2}$	$\frac{2}{t}(1-\cosh at)$
$\arctan\frac{k}{s}$	$\frac{1}{t}\sin kt$
$\frac{1}{s^{\alpha}}$	$\frac{t^{\alpha-1}}{\Gamma(\alpha)}$
$\frac{s^{\alpha-1}}{s^{\alpha}\mp\lambda}, \Re(s) > \|\lambda\|^{1/\alpha}$	$E_{\alpha,1}(\pm\lambda t^{\alpha})$
$\frac{k!s^{\alpha-\beta}}{(s^{\alpha}\mp\lambda)^{k+1}}, \Re(s) > \|\lambda\|^{1/\alpha}$	$\varepsilon_k(t,\pm\lambda;\alpha,\beta)$
$\frac{k!}{(\sqrt{s}\mp\lambda)^{k+1}}, \Re(s) > \lambda^2$	$t^{\frac{k-1}{2}}E^{(k)}_{\frac{1}{2},\frac{1}{2}}(\pm\lambda\sqrt{t})$

Glossary

Fractional calculus is a branch of mathematical analysis that studies the possibility of differentiation and integration of arbitrary real or complex orders of the differential operator.

Differintegral in fractional calculus is a combined differentiation/integration operator.

Short memory principle means taking into account the behavior of function only in the “recent past”.

Singlevalued function is a function that, for each point in the domain, has a unique value in the range.

Multivalued function is a function that assumes two or more distinct values in its range for at least one point in its domain.

State space is the set of all possible states of a dynamical system, where each state of the system corresponds to a unique point in the state space.

Equilibrium point, sometimes called fixed point is a solution of the autonomous system of ordinary differential equations that does not change with time.

Jacobian matrix is the matrix of all first-order partial derivatives of a vector-valued function. It describes the orientation of a tangent plane to the function at a given point. The behavior of the system near a stationary point is related to the eigenvalues of the Jacobian at the equilibrium point.

Saddle point is when all eigenvalues are real and at least one of them is positive and at least one is negative. Saddles are always unstable.

Node point is when all eigenvalues are real and have the same sign. The node is stable (unstable) when the eigenvalues are negative (positive).

Focus point, called also spiral point, is when eigenvalues are complex-conjugate. The focus is stable when the eigenvalues have negative real part and unstable when they have positive real part.

Focus-Node point is when it has one real eigenvalue and a pair of complex-conjugate eigenvalues, and all eigenvalues have real parts of the same sign. This equilibrium is stable (unstable) when the sign is negative (positive).

Saddle-Focus point is when it has one real eigenvalue with the sign opposite to the sign of the real part of a pair of complex-conjugate eigenvalues. This type of equilibrium is always unstable.

Center equilibrium occurs when a system has only two eigenvalues on the imaginary axis, namely, one pair of pure-imaginary eigenvalues.

Instability measure for equilibrium points of the fractional-order nonlinear system is mathematically equivalent to difference between the angle of stability border given by necessary stability condition and minimal angle of all roots obtained from characteristic equation of the system.

Lyapunov exponent of a dynamical system is a quantity that characterizes the rate of separation of infinitesimally close trajectories.

Index

MIX
Papier aus verantwortungsvollen Quellen
Paper from responsible sources
FSC® C105338

If you have any concerns about our products,
you can contact us on
ProductSafety@springernature.com

In case Publisher is established outside the EU,
the EU authorized representative is:
Springer Nature Customer Service Center GmbH
Europaplatz 3, 69115 Heidelberg, Germany

Printed by Libri Plureos GmbH
in Hamburg, Germany